# A PRACTICAL GUIDE TO INSPECTING

# STRUCTURE

*By Roy Newcomer*

# CONTENTS

# INTRODUCTION

My background includes many years in construction and several more as the owner of a Century 21 real estate franchise. In 1989, I started a home inspection company that grew larger than I ever imagined it could. Training my own staff of inspectors to the highest inspection standards led to my teaching home inspection seminars across the country and developing study courses, books, and videos for home inspectors. The American Home Inspectors Training Institute was founded as a result of my desire to share this experience and knowledge in home inspection.

The *Practical Guide to Inspecting* series is intended for both beginning and experienced home inspectors. So if you're studying home inspection for the first time or are using the materials as a refresher, these guides should be of assistance to you.

I've created these guides to include all aspects of home inspection. Not only a broad technical background in home systems, but the other things you need to know in order to perform a *good* inspection of those systems. They lay out technical information, guidelines for the inspection, how-to instructions for inspecting system components, and the defects, deficiencies, and problems you'll be looking for during the inspection. I've also included some advice on how to report your findings to the home inspection customer.

I've been a member of several professional organizations for a number of years, including ASHI® (American Society of Home Inspectors), NAHI™ (National Association of Home Inspectors), and CREIA® (California Real Estate Inspection Association). I am a great supporter of those organizations' quest to promote excellence in home inspection.

I encourage you to follow the standards of the organization to which you might belong, or any state regulation that might take precedent over the standards used here. Use the standards in this book as a general guide for study and apply the standard or state regulation that applies to you.

The inspection guidelines presented in the Practical Guides are an attempt to meet or exceed standards and regulations as they exist at the revision date of the guides.

There's a lot to learn about home inspection. For beginning inspectors, there are some *hands-on exercises* in this guide that should be done. I'm a great believer in learning by doing, and I hope you'll try them. There are also some of my *personal inspection stories* to let you know what it's really like out there.

The *inspection photos* referenced in this text can also be found on www.ahit.com/photos. You'll read the story about each one as you go along. Be sure to watch for my *Don't Ever Miss* lists. I've included them to alert home inspectors to report those defects (if found during the inspection) in the inspection report. If missed, these items are often the cause for lawsuits later. Finally, to help you see how you're doing as you study this guide, I've included some *worksheets*. The answers are given for each one for self checking. Give them a try. Checking yourself can help you lock important information in your mind. There's also a *final exam* that you can complete and send in to us. Many organizations and states have approved this book for continuing education credits. Submit the exam with the required fee if you need these credits.

In total, the *Practial Guide to Inspecting* series covers all aspects of the general home inspection. Each guide covers a major aspect of the inspection, as their titles show:

*Electrical*
*Exteriors*
*Heating and Cooling*
*Interiors, Insulation, Ventilation*
*Plumbing*
*Roofs*
*Structure*

If you are interested in other titles in the series, please call us at the American Home Inspectors Training Institute to order them. Call toll free at 1-800-441-9411.

*Roy Newcomer*

# INSPECTING
# STRUCTURE

The *inspection photos* referenced in this text can also be found on www.ahit.com/photos. You'll read the story about each one as you go along. Be sure to watch for my *Don't Ever Miss* lists. I've included them to alert home inspectors to report those defects (if found during the inspection) in the inspection report. If missed, these items are often the cause for lawsuits later. Finally, to help you see how you're doing as you study this guide, I've included some *worksheets*. The answers are given for each one for self checking. Give them a try. Checking yourself can help you lock important information in your mind. There's also a *final exam* that you can complete and send in to us. Many organizations and states have approved this book for continuing education credits. Submit the exam with the required fee if you need these credits.

In total, the *Practial Guide to Inspecting* series covers all aspects of the general home inspection. Each guide covers a major aspect of the inspection, as their titles show:

*Electrical*
*Exteriors*
*Heating and Cooling*
*Interiors, Insulation, Ventilation*
*Plumbing*
*Roofs*
*Structure*

If you are interested in other titles in the series, please call us at the American Home Inspectors Training Institute to order them. Call toll free at 1-800-441-9411.

*Roy Newcomer*

# INSPECTING STRUCTURE

# Chapter One

# THE STRUCTURAL INSPECTION

The structure of a home is its **skeleton**, including the foundation and footings, the roof, and the framework — the floor, wall, and ceiling structures. Think of the home stripped of its exterior and interior finishings, standing naked without its siding and roofing materials and the interior wall, floor, and ceiling coverings. You would see exposed studs, beams, rafters, joists, and other structural components of the home.

The structure of a home is not seen in its entirety except during the construction process. Once the home is finished, much of its structure is **buried below the ground** or **hidden behind the coverings**. The home inspector learns to inspect these hidden structural components by looking at clues to what is going on under the surface. Surface cracks can indicate that the structure is moving or breaking apart. Sagging and buckling can indicate failing structural members. Staining on surface finishes may signal water penetration into the unseen components of the structure. The home inspector learns to read the visible signs.

But even with excellent training and structural knowledge, the home inspector is not always able to determine whether structural defects are present. Sometimes there are no external signs of what's going on behind the scene. Exterior and interior finishes, as well as repair and patching work can conceal some problems with the actual structure of the house. Sometimes the identification of problems is just not possible. When problems are suspected but unable to be identified, the home inspector may recommend that a structural engineer be consulted.

Inspecting the structure of the home is the **most important step** in the home inspection process. The general usefulness of the whole home depends on how sound its structure is. And, in cases of extreme structural defects, the safety of the occupants is at risk. The home inspector observes all those aspects of the home's structure that are readily available for inspection, describes the type of construction and the materials used in the structural components, and reports any problems or defects found. At the same time, the inspector is dealing with major questions about the home's structural integrity such as the failure and collapse of any structural system.

*Guide Note*

*Pages 1 to 4 lay out the content and scope of the inspection of the structural components. It's an overview of the inspection of the home's structure, including what to observe, what to describe, and what specific actions to take. Now's the time to study these requirements in depth.*

*Definition*

*The  structure of a home is its skeleton, including the foundation and footings, the roof, and the framework.*

INSPECTING
STRUCTURE

- Foundation
- Floor structure
- Wall structure
- Columns
- Ceiling structure
- Roof structure

## Inspection Guidelines and Overview

Read through and study these standards of practice carefully. Pay attention to the stated objective of the structural inspection.

| OBJECTIVE | To identify major deficiencies in the condition of structural members which affect their load bearing capacity, and major deficiencies in the structural assembly or sub-assemblies. |
|---|---|
| OBSERVATION | Required to inspect and report:<br>• Foundations<br>• Floors<br>• Walls<br>• Columns<br>• Ceilings<br>• Roofs |
| ACTION | Required to:<br>• Probe structural components where deterioration is suspected.<br>• Enter underfloor crawl spaces and attic spaces except when access is obstructed, when entry could damage property, or when dangerous or adverse situations are suspected.<br>• Report the methods used to observe underfloor crawl spaces and attics.<br>• Report signs of water penetration into the building or signs of abnormal or harmful condensation on building components.<br>• Report the presence or absence of seismic anchoring and bracing components, where applicable. |

Not every detail of what is to be inspected and what is to be reported are stated in these standards. This is just an outline of the structural inspection. There are many other details that you'll learn in this guide. But for now, this is an overview of the inspection for each of the major structural components listed in the chart above.

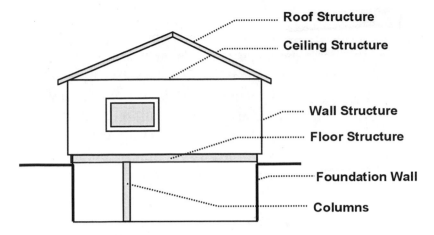

Roof Structure

Ceiling Structure

Wall Structure

Floor Structure

Foundation Wall

Columns

*Footings* are the bases on which the foundation rests. Footings support and distribute the weight of the structure to the soil

The *foundation* is that part of a structure that supports it, transmits the weight of the structure from above-grade walls to the footings, and protects the structure from the effects of soil pressure upon it.

**Foundations:** The home inspector examines the foundation walls from outside and inside, where possible. The inspector is required to describe the **materials** and **types of foundation construction**. Examples of materials used in foundations are stone masonry, brick, poured concrete, concrete block, and wood. Examples of foundation structures are basements, crawl spaces, slab-on-grade, and piers or posts.

The inspector examines the **condition** of the foundation walls and reports any **defects** found — deteriorating materials, cracking, movement, bowing, and water penetration, for example. The inspector is also looking for any defects in the foundation **footings**, which are usually below the ground and not visible. Footing movement or failure will evidence itself above ground in the cracking or displacement of the foundation wall and the exterior walls.

**Basements and crawl spaces:** The foundation walls are also inspected from inside the basement. The inspector is required to report signs of **water penetration** into the building and pays attention to basement moisture and provisions for drainage.

Note that the inspector is required to **enter the underfloor crawl space** whenever possible and to report the methods used to observe it. The inspector should make every effort to get into the crawl space. The more difficult the access is, the more likely it is that problems are present. Reporting where the crawl space was viewed from and/or whether the crawl space was not observed is for the inspector's protection.

*Columns* are vertical supports that carry the weight of the structure from the girders (or beams) to the ground. Columns transmit weight to footings.

*Girders* (or beams) are horizontal load bearing members of a floor system that carry the weight of the floor and wall loads to the foundation and columns.

*Joists* are horizontal members of a floor system that carry the weight of the floor to the foundation, girders, or load-bearing walls.

The *subflooring* transfers the load of the home's furnishings and people to the floor joists

**Guide Note**

*Ceiling, wall, and floor finishes are not part of the structural inspection. That's presented in another of our guides —A Practical Guide to Inspecting Interiors, Insulation, Ventilation.*

**Columns:** The inspector examines the supports (columns and posts) inside the foundation. Columns are generally inspected from the basement or crawl space. The inspector describes the type of materials used and inspects for any deterioration in these materials such as rust on steel columns and rot in wood posts. Note that wooden structural members are **probed** only when deterioration is suspected. The inspector uses the tip of a screwdriver, ice pick, or knife to test for wood rot.

**Floors:** The inspector describes the materials and condition of the floor in the basement or crawl space. The floor system of the above-grade part of the structure is usually only visible from underneath, from the basement or crawl space. The inspector describes the materials and condition of girders, joists, and the subfloor visible from below. The inspector inspects the junctions of the members for structural integrity — where girders rest on the foundation and where joists meet above girders — and checks the floor joists for proper span.

**Walls:** In the inspection of the structural components of the house, the inspector identifies the wall construction — whether solid masonry walls, wood frame, log, or post and beam. The wall finishes in the interior of the home are not part of the inspection of structure. Here, the inspector is concerned with what the construction is and the condition of the visible aspects of exterior wall construction and gross cracking or displacement of interior walls.

**Ceilings:** Ceiling structure is visible from the basement where the inspector makes the assumption that the methods of construction at the basement ceiling are used throughout the structure. The top-most ceiling in the structure may be visible for inspection from the attic.

**Roofs:** The inspector gets into the attic, where possible, to observe the roof structure. Sometimes, that's not possible. Again, as with crawl spaces, it's important for the inspector to record the method of observation — whether from inside the attic itself or from pull-down stairs. The inspector describes the type of roof structure, whether rafters or trusses, and inspects structural members for condition and structural integrity.

# Chapter Two

# THE DYNAMICS OF STRUCTURE

The structure of a house is more than its foundation, framework, and roof. It's a dynamic system, designed to protect and preserve itself and the interior spaces it encloses from the forces working against it. Think of the structure as a fortress holding back the enemy. You might think the structure is constructed, the finishing coverings added, people move in, and that's that. Not so. The structure of a home is in a continual battle with elemental forces.

## Weight

A structure is expected to stand up. The force of **gravity** is constantly working to bring it down. A structure has to be designed to resist gravity. It must support its own weight, which is called the **dead load** of the structure. It must also support the **live load** of the structure, which includes the people inside, the furniture, and other weight such as snow on the roof. A structure *does* transfer its weight to the soil — as gravity would have it — but in a well-designed way. If the skeleton of the structure is weak, it cannot support the dead and live loads of the home, and the structure can collapse.

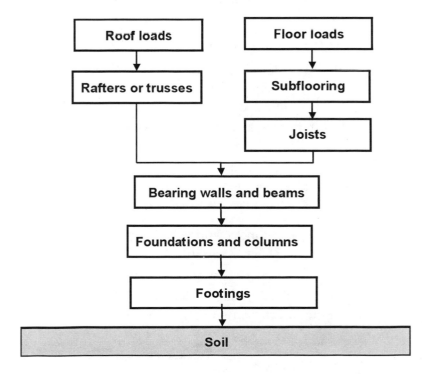

*Guide Note*

*Pages 5 to 8 present a discussion of the forces that act on the structure of a building.*

*Definitions*

*The dead load of a structure is the weight of the structure itself and its sheathing and wall-coverings, and other integral components.*

*The live load is the weight of the home's occupants, the furnishings of the home, and other weight that the structure must support.*

## Movement

Not only is a structure expected to stand *up*, a structure is expected to stand *still*. Whether a structure moves or not is partially determined by the condition of the soil underneath and around it. The house must be sitting on soil strong enough to support it. When the ground below the house fails, the house sinks. Picking a good site is an important step in construction.

Soil is a dynamic system. Soil moves. A house at the bottom of a hill can have soil moving toward it over time and exerting more and more pressure against the foundation. Such a house may be pushed from its original position. And a house built at the edge of a bank may begin moving toward the edge of the bank as soil erodes away.

The foundation resists the **pressure of the soil** against the structure, even in homes built on the proper site. The foundation must be built strong enough to resist this natural pressure. In earthquake zones, structures must follow local building codes to resist violent earth movement.

## Wind and Water

A structure is expected to *hold together* under the forces that act upon it. Wind acts upon the structure from time to time and in varying degrees — from breezes up to hurricane forces. Wind can push and pull at a building and also try to lift it. Structures must be strong enough to resist this pressure.

Although water doesn't appear to have the power of other forces such as strong winds, it poses a big challenge to structures. Water plays a large role in the action of the soil upon the foundation of the structure. Increasing the water content of the soil increases its pressure against the foundation. Then as the soil dries, it can shrink away from the foundation. Frozen water in the soil also causes it to expand and exert increased pressure against the foundation from the sides and from below.

Underground water can move soil away, creating voids under a structure, causing the structure to settle. It will also try to enter the structure. The materials used in construction must be able to protect the structure against water penetration and the resulting deterioration of those materials. The structure must provide the proper drainage systems to deal with underground water.

## Internal Stresses

The integrity of a structure depends on each of its individual structural members. A structure can be said to be in a constant state of stress as individual members exert forces upon each other. For example, floor joists spanning the structure would sag without the restraint and support provided by the girders, which in turn push on supporting columns. These members must work in concert with each other without damaging each other.

Two forces that individual structural members experience are **compression** and **tension**. Compression pushes at building components; tension pulls at them.

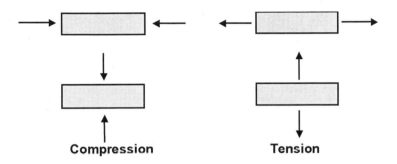

Compression | Tension

Structural members under compression tend to get shorter; those under tension longer. Some members may be under both stresses. For example, collar ties withstand compression loads **along the grain** of the wood from the weight of the roof pushing the rafters inward. They could experience tension loads from the outward push of the rafters. Girders withstand compression **against the grain** from joists above and columns below.

Two other stresses the structure must be able to withstand are **shearing** and **bending**. Individual structural members may fail under these stresses. Shearing occurs when nonaligned forces are applied against a member from opposite directions, causing the material to tear.

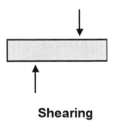

Shearing

Movement of a structural member out of its original position without shearing is called bending. Such members can sag and buckle from the forces applied to them.

Materials used in a structure are chosen for their ability to withstand the internal stresses of the structure. Some materials will do well, while others won't.

### Definitions

*Compression* is a stress that pushes on a structural member, tending to make the member reduced in size.

*Tension* is a stress that pulls at a structural member, tending to make the member increased in size.

*Shear* is a stress resulting from forces being applied upon a structural member from opposite directions. *Shearing* can cause the structural member to crack or split or completely separate.

*Bending*, in terms of structure, refers to the movement of a structural member out of its original position without shearing as a result of forces applied to the member.

## Thinking about Structure

The home inspector must understand the dynamics of structure and should be thinking about the forces acting both against and within the structure. The image of the **fortress** standing firm against and doing battle with these "enemies" may be an overstated, dramatic image, but it's not far from the truth. The structural integrity of a home must be sound and strong.

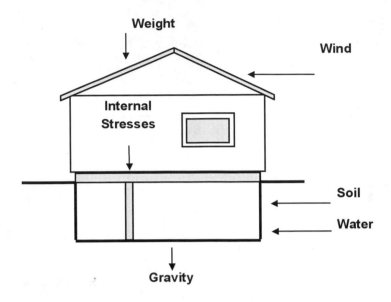

The home inspector, viewing a house from the structural standpoint, should be thinking about gravity and weight, the soil, water and wind, and the actions of individual members upon each other. These are the **causes** of what can go wrong with the structure. Understanding these causes helps the home inspector recognize and understand their **effects** upon the structure.

- Movement of the entire structure or parts of it
- Distortions of the structure such as leaning, bowing, sagging, and buckling
- Failure or defects of any structural components as evidenced by cracking and splitting
- Water penetration
- Deterioration of building materials

NOTE: Not to be forgotten in the discussion of structural dynamics is the use of the **right building materials** for each structural member and the **proper construction methods**. One without the other will damage the structure. The lack of both can be a total disaster. This guide will address both issues.

# Chapter Three

# FOUNDATION CONSTRUCTION

This chapter will present the various types of foundation construction.

## Footings

Footings support and transmit the **dead load** weight of the structure itself) and the structure's **live load** (weight of its occupants and furnishings) to the soil. Footings are found below foundation walls and under columns or posts. Their purpose is to spread these dead and live weights over a large area and prevent the structure from sinking into the soil.

Footings should be built **below grade** on undisturbed soil and be wide enough so the loading per square foot is less than the bearing capacity of the soil. Footings in cold climates are laid **below the frost line** to keep the foundation and columns from lifting if moisture in the ground freezes. Because footings are laid below grade, they are not usually visible. Footing materials can be stone, brick, or

**Foundation**

**Slab**

**Footing**

**Post**

**Footing**

### Guide Note

Pages 9 through 18 present a discussion on foundation and slab construction.

### Definitions

*Grade* is the slope of the land at the building site. *Below grade* means below the surface of the ground.

The *frost line* marks the depth of penetration of frost into the ground. In cold climates, the frost line is normally about 4' below the grade.

**#1 Footing for new foundation**

***Photo #1*** shows the **footing** laid for a new foundation. This footing is poured concrete, poured as a continuous form at the bottom of the excavation. The footing is put in first at the site. The next step is for the construction crew to start laying the block or pouring the concrete for the foundation.

## Definitions

*A <u>footing drain</u> is a drainage system laid around the perimeter of a foundation below the level of the slab which drains water from the soil to another location. The <u>drain tile</u> used in footing drains today is made of flexible, perforated plastic piping.*

*A <u>sump</u> is a pit located under the basement floor and contains an electric pump. The pit collects water from the footing drains and the <u>sump pump</u> pumps it away from the house.*

concrete. Today, however, almost all footings are **poured concrete**. They are normally from 16" to 24" wide and from 6" to 16" deep.

In those areas where soil conditions prevent the use of footings, then pilings, grade beams, and floating slabs may be used instead of footings.

## Footing Drains

Foundation moisture exposure is by footing drains where water above or below grade is present and poses a problem. **Drain tile** is laid around the perimeter of the foundation next to the footing and below the level of the floor slab. The use of perimeter drainage systems started after World War I, and at that time the drain tile was 4" clay tile pipe that came in sections.

Today, a flexible and perforated plastic piping is used. The drain tile is laid, perforations down, around the perimeter of the foundation at the footing level. Filter paper is laid above and below the tile to prevent the drain tile from clogging. Then the tile is covered with at least 6" of gravel or crushed stone.

Some areas of the country require both **exterior** and **interior** drain tile. Some require only exterior drain tile. The requirements vary.

The drainage system collects water and drains it into a sewer system (where available), or away from the house to emerge above ground (if the slope permits), into a drywell or underground gravel pit at least 15' from the building, or inside the foundation perimeter to a **sump**, which pumps the water out to a sewer system or to some distance from the house.

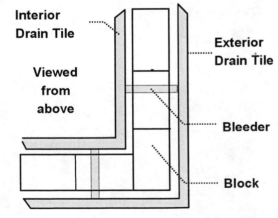

A sump is a pit located below the basement floor. It can have concrete or earth floors and walls. Plastic liners may be used. An electric **sump pump** is located in the pit to pump water coming from the perimeter drainage system away from the house. A **bleeder** runs from the exterior drain tile through or under the footings to interior tile, if present, and to the sump.

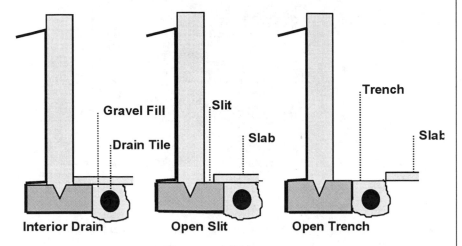

**Interior footing drains** may be added underneath the basement floor later to deal with a water problem — perhaps after the exterior drain tile has failed. Putting in interior drains to solve the problem is less expensive than exterior work. A trench is usually dug, breaking through the concrete floor around the inside of the foundation wall. Drain tile and gravel fill is laid so water can run to a sewer system or to a sump. Concrete may be poured to cover the drains, but you may also see an open trench or slit left in the basement floor.

**#2 Drain tile at builiding site**

*Photo #2 shows drain tile at the building site. The drain tile is the black flexible piping in the photo.*

**#3 Drain tile in position**

*Photo #3 shows it in position and covered with a cloth "sock" to prevend soil from clogging pipe. Once this site is backfilled, the drain tiles will not be visible*

## Foundation Walls

The foundation is that part of the structure that supports it, transmits the weight of the structure from above-grade walls to the footings, and protects the structure from the effects of soil pressure upon it. The foundation wall, which carries this weight below the frost line, sits on top of the footing.

Because soil pressure increases with depth, a foundation wall below grade must be stronger the deeper it goes. Local codes and regulations will set the proper thickness for various types of foundations at various depths. For example, concrete block walls need to be 8" thick at a depth of 5' but 12" thick at a depth of 7', whereas poured concrete foundations need to be only 6" and 8" thick at those same depths.

## The Block Foundation

The **concrete masonry unit** (CMU) describes various kinds of hollow-core blocks used in foundation construction. **Cinder block** is made from slag from steel making or cinders from the railroad. Cinders and slag are no longer widely used as components, however they enjoyed a long history in home construction and generally did well. However, both types of cinder blocks can deteriorate over time. Both can retain moisture and can break from freezing and thawing. The iron particles in slag can expand when they rust and cause spalling and staining on the face of the block.

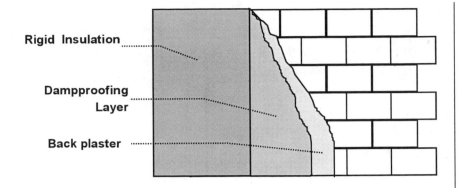

Rigid Insulation

Dampproofing Layer

Back plaster

**FOUNDATION TYPES**

- Concrete masonry unit
- Poured concrete
- Stone masonry
- Brick
- Wood foundation
- Piles, piers, and grade beams

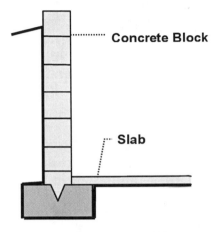

Concrete Block

Slab

**Concrete block** is made from crushed stone and builder's sand. Many foundation walls are made with concrete block. Concrete blocks are **mortared** and can be reinforced with wire mesh laid in the mortar every three or four courses and/or with vertical steel bars held in place inside the block cavity with grout. Concrete block foundations are normally **back-plastered** (parged) and a bituminous or plastic **dampproofing coat** is applied for waterproofing. Take another look at **Photo #2** which shows the concrete block in the process of going up. Notice that some back plastering has been started on the walls.

**#4 Concrete block foundation**

*Photo #4 shows the exterior of a concrete block foundation with **exterior insulation** which is recommended by experts. This insulation is a rigid board made of styrene or urethane foamed material. This type of insulation lets water flow down to the drain tile and helps keep the basement dry.*

## Poured Concrete

In the poured concrete foundation wall, **forms** are set up on the footings and concrete is poured into them. Bars and wire reinforcing can be erected in the forms before the concrete is poured. To hold the forms apart and support them, wire or metal bar ties (or wall ties) may be run from side to side within the forms and left in the concrete after it has cured. Sometimes metal clips are used to tie the forms together where they meet. The visual parts are removed after the pour. The clips leave vertical indentations or seams in the surface of the wall.

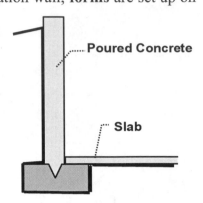

Poured Concrete

Slab

The ends of any iron reinforcing members exposed at the top of the foundation after the concrete has set are usually removed and the hole left filled with mortar. This is done to protect the iron left in the concrete from rusting and damaging the wall.

The poured concrete foundation, using high-quality concrete and poured and left to cure properly, provides a very strong wall. However, the job can be botched. Failures of the wall can result from the wrong formulation of concrete, pouring in temperatures too hot or below freezing, or removing the forms as soon as the concrete sets. Concrete needs time to cure. Poorly cured concrete can be soft and powdery. When this happens, there can be **spalling** on both the exterior and interior faces of the wall. This is where the surface of the concrete crumbles. A really bad job can cause the wall to collapse. Letting one section set up before another is poured can cause cold joints in the surface.

#5 Poured concrete foundation

*Photo #5 shows a poured concrete foundation. The forms used here impressed a brick look to the surface of the interior walls. Notice the vertical form lines.*

## Stone Masonry

A stone masonry foundation can be either **dry laid** or **laid in mortar**. The dry laid stone wall was built with no mortar and with no footings. Gravity held the piled stones together. The loads these walls support are distributed from stone to stone, from point to point. Smaller stones may have been inserted into gaps. These stones are usually non-structural, and are not load bearing. They may even fall out without damage to the structural integrity of the stone foundation.

Dry laid masonry walls, without the mortar seal, can let in water, soil, and air. Sometimes these old foundations have been treated in some way to eliminate water problems. Concrete grout can be poured between and around the stones to absorb the external soil pressure. Sealing the interior of the wall with repointing or waterproofing is not usually helpful because water still enters the wall and pushes against the finish.

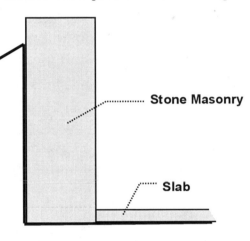

Stone Masonry

Slab

The stone foundation is usually very thick and can last forever. It can be constructed with fieldstone, limestone, or other types of stone used over 100 years ago.

Stone foundations laid in mortar, have loads transmitted both through the mortar and the stones. In good form, these walls are solid and waterproof. However, the lime mortar used by masons in earlier times can deteriorate due to exposure to humidity, soil moisture, and condensation over time. The mortar actually turns into sand and falls out. Because the mortar originally used strengthened the wall, the mason may have used stone of a lower quality than the dry laid foundation. Without the support of the mortar, the wall acts like it's dry laid and the lower-quality stones can crack and crush.

*Definitions*

*The dry laid stone foundation is a wall constructed of stone piled on stone without mortar and with no footings.*
*Mortar is a mixture of a binder (lime, masonry cement, portland cement), an aggregate (sand), and water in many variations. It is used to bond masonry units such as concrete blocks, stones, and bricks together.*

*For Beginning Inspectors*

*If stone foundation walls are common in your area, you might want to visit a mason who's repaired some of them. Ask what particular problems the mason has seen in walls of this type.*

## Brick Foundations

The brick foundation is less common than other types and is quite uncommon in some parts of the country. Some very old brick foundations may still be in very good condition. However, a great variety of bricks were put into these foundations, made from different clays and additives. Some bricks are appropriate for below grade and others are not.

Bricks are strong under compression but weak and brittle under tension. Bricks are also porous; they actually pull water from the soil into themselves. This process can hasten the deterioration of the mortar in the brick foundation. Salts in the ground water may cause the brick to deteriorate.

On the outside of the house where the brick foundation is above grade, rain and roof runoff may wash against the brick. Mortar can be washed away and bricks fall out. Water inside the bricks can cause spalling, where the surface of the brick crumbles away. When water freezes inside the brick, you may see the surface of the brick coming off in thin sheets.

## Wood Foundations

Wood foundations are relatively new, but reports to date from builders and homeowners have been positive. A permanent wood foundation offers a superior drainage system and wood stud interior walls that can be insulated and finished off as other interior walls.

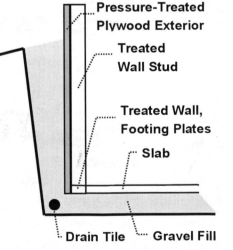

A wood foundation is basically a wall of studs and plywood sheathing, built entirely of **pressure-treated wood**, impregnated with chemical preservatives. The chemicals protect the wood against attacks of termites and the fungi that cause rot.

The basement walls and floor rest on a bed of gravel, with more gravel backfilled around the foundation. It isn't always

**#6 Wood foundation**

*Photo #6* shows a *wood foundation* put up at a site. Notice the prepared bed of gravel and the floor plate at the bottom of the wall. The cross pieces halfway up the wall are blocking between the studs.

easy to recognize a wood foundation because the interior is likely to be covered with drywall or other covering. One clue is that the floor slab is poured against the wood foundation wall plates to hold them in place against the pressure of the backfill.

From the exterior, you'll usually find that there is some sort of protective and decorative covering applied over the portion of the foundation above grade.

The slab floor with cracks, tilting, humps, or hollows can be a sign that the wood foundation wall is failing or the drainage system isn't working. The smell of decaying wood is a bad sign.

When inspecting the wood foundation, look under stairwells or other work areas where walls may not be covered. If you suspect problems that aren't visible, the only solution may be having the owner remove the wall covering so you can take a look at the foundation.

## Piles, Piers, and Grade Beams

A foundation doesn't have to be a continuous form or wall. Piles and piers are columns that serve the same purpose as the foundation wall by supporting the rest of the structure and transmitting the weight of the structure to the footings or to the soil.

**Piles** are generally steel, wood, or concrete. They're used when the soil can't support the load of the structure. Piles are actually driven into the ground by a machine to reach a soil of bearing strength or until friction prevents any further movement.

The top of each pile is topped with a concrete cap. This cap acts as a girder upon which the rest of the structure is built.

**Piers** are columns of stone, brick, concrete block, cinder block, or wood. Piers can be 12" to 14" in diameter and from 6' to 30' in length. They're built in a hole which is then filled in around them. Piers should sit on footings below the frost line. Piers are also topped with a concrete cap. A crawl space usually has a pier system on which the house is built. You don't often see piers used when there is a basement.

**Grade beams** are poured reinforced concrete beams that rest on the grade or just in the ground or that rest on piers with the cap in or above the ground. Some garages are built on grade beams supported by piles. You may find grade beams in houses under perimeter or load-bearing walls.

*Photo #7 shows a **ICF** foundation – Insulated concrete forms. These insulated forms can be used for the foundation or complete exterior walls.*

**#7 ICF Foundation**

**#8 Pre-Cast Foundation**

*Photo #8 shows a **Pre-Cast Foundation**–This is a prefabricated foundation system, consisting of steel reinforced, insulated concrete panels that are trucked to the site and set in place with a crane.*

# Chapter Four

# SLAB CONSTRUCTION

In **slab-on-grade** construction, a poured concrete slab rests directly on the ground without room for a basement or crawl space underneath. The house is built on the slab, which supports the structures above and acts like a footing for it. Slabs are generally about 4" to 6" thick. In some types of slab-on-grade construction, the edges are made thicker, and the slab may be strengthened under interior load-bearing walls, chimneys, and fireplaces. The slab may rest on a conventional foundation.

A slab may be reinforced with wire mesh to prevent shrinkage cracks. Heating ducts may be incorporated into the slab, but conduit for electrical and telephone wiring is generally not laid under the slab. Natural gas piping must be run overhead unless properly sleeved beneath the slab. In some areas, local codes forbid any openings in the slab such as openings for pipes and other utilities. Under-slab water supply piping must be laid before the concrete is poured.

The slab should rest on compacted fill of clean soil, sand, or gravel and have a **vapor barrier** between the bottom of the slab and the fill.

In the **monolithic slab-on-grade**, as shown here, the slab and the foundation are poured as one piece. Reinforcing wire mesh, steel bars, or post-tension cables can be used to strengthen concrete between the slab and its thicker edge. Tubes or tunnels for the heating system can be incorporated into the monolithic slab. A vapor barrier such as sheets of polyethylene film separate the slab from the soil surrounding it. Rigid insulation can be added around the perimeter.

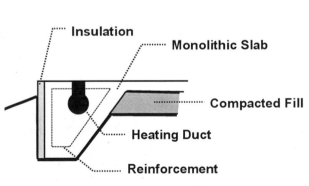

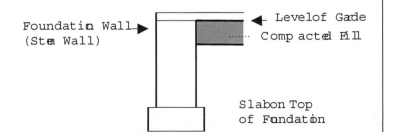

---

**Slab-on-Grade**

- Monolithic
- Resting on foundation
- Supported by the foundation
- Floating and pinned to foundation
- Post-Tensioned

*Definitions*

*Slab-on-grade* construction consists of a poured concrete slab that rests directly on the ground.

*In the monolithic slab-on-grade, the slab and foundation are poured as one piece.*

*The <u>resting slab</u> is laid to rest on top of a conventional foundation.*

*The edges of a <u>supported slab</u> rest on a ledge at the top of the foundation wall.*

*An independent or <u>floating slab</u> does not rest on the foundation at all.*

Slab-on-grade construction may make use of the conventional foundation and footing, which may be only a foot deep or extend down below the frost line. The slab may be poured to **rest on the foundation**, as shown here. This type of slab-on-grade method also requires the slab to rest on well-compacted soil and have a vapor barrier under the slab.

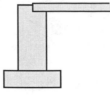

**Slab on Shoulder of Foundation**

The slab can be **supported by the foundation** as shown in the picture on the left. Just like the slab that rests on the foundation, this slab is not poured at the same time as the foundation. The footing and foundation is poured first with a 4" ledge at the top of the foundation wall. Then the slab is poured. This ledge supports the slab at its perimeter.

A slab can be installed that is basically independent of the foundation or **floating** on the soil. The picture on the right is an example. In this case, the slab is sometimes pinned to the foundation with the use of steel pins or dowels.

**Slab Pinned to Foundation**

Another relatively new slab in use is the post-tensioned type. This is a monolithic slab that utilizes cables within the slab, which are tensioned after the slab cures. The intent is to compress the concrete and hold the entire slab together as one piece. This reduces structural issues if there is any underlying settlement or expansion of the soil.

In a house built on a slab, it's not easy to determine the design of the slab. Its underside is not visible, the foundation is not accessible, and there may be wood flooring laid over the slab. We'll talk more about this later in the guide.

#9 Slab already poured

*Photo #9 shows a **slab already poured**. Notice the plumbing pipes sticking out from the slab.*

# WORKSHEET

*Test yourself on the following questions.*
*Answers appear on page 22.*

1.  Identify the type of foundation shown in Photo #5.

    A.  Cinder block
    B.  Brick
    C.  Poured concrete
    D.  Monolithic slab

2.  What part of a home's structure is <u>least likely</u> to be visible to the home inspector?

    A.  Footings
    B.  Foundation
    C.  Crawl space
    D.  Attic

3.  How does compression act on structural members?

    A.  It pulls at the members and makes them longer.
    B.  It pushes at the members and makes them shorter.

4.  When should the home inspector <u>not</u> enter the crawl space?

    A.  When access is obstructed.
    B.  When entry could damage property.
    C.  When dangerous or adverse situations are suspected.
    D.  All of the above.

5.  In Photo #3, what is pictured on the exterior of the poured concrete wall?

    A.  Damp proofing
    B.  Shearing
    C.  Insulation
    D.  Back plastering

6.  What is the white object at the base of the wall in Photo #3?

    A.  Drain tile
    B.  Heating duct
    C.  Footing
    D.  Dampproofing

7.  Which force would exert the <u>least</u> amount of stress on a foundation?

    A.  Wind
    B.  Water in the soil
    C.  Soil pressure
    D.  Gravity

8.  Which of the following drawings show the independent floating slab?

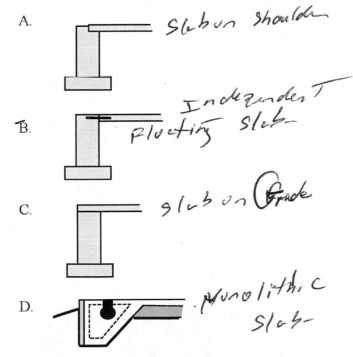

    A.     Slab on Shoulder
    B.     Independent Floating Slab
    C.     Slab on Grade
    D.     Monolithic Slab

9.  *Think about this one.* Why do you suppose that all guidelines require the home inspector to report the methods used to observe underground crawl spaces and attics?

Pages 22 to 42 present the procedures for foundation inspections.

**Worksheet Answers** *(page 21)*
1. *C*
2. *A*
3. *B*
4. *D*
5. *A*
6. *A*
7. *A*
8. *B*
9. *Because defects can be present in a crawl space or attic, the home inspector must let the customer know if the inspector was unable to gain access to the crawl space or attic to find those defects. The inspection guidelines stress this point because of the liability. The customer should understand that defects cannot be reported if access is impossible or limited.*

# Chapter Five

# INSPECTING FOUNDATIONS

The home inspector starts the inspection of the foundation from the **outside** of the house, watching for signs of footing and foundation failure on the exterior. The inspector looks for **settlement or movement** of the house and their effect on the rest of the structure. While some signs note obvious problems, other signs may only indicate that the inspector must continue to investigate to find the cause of the exterior sign. The inspector focuses on the following:

- The ridge line of the roof
- Racking or leaning of the house
- The chimney pulling away from the house
- Twisted siding
- Movement of foundation wall and cracks
- Corners of the building settling
- Cracks in exterior walls
- Displacement of windows

The inspector also inspects the foundation from **inside,** from the basement or crawl space, paying attention to:

- Cracks in foundation walls
- Bowing or leaning of foundation walls
- Window displacement
- Deterioration of materials in the walls, piers, and supports
- Evidence of water penetration

## Cracks

Wall cracks appear as the result of overloading or because of settlement or heaving. The location and pattern of the cracks can be a clue to what is going on with a foundation.

**Vertical cracks** are most often caused as a result of settlement of the structure, soil compacting, or soil washing away under the footings. It occurs when there is an upwards overload force next to a downwards overload force. They can also occur after remodeling, when new dead loads are added to the structure, or by adding very heavy live loads.

Vertical cracks that extend down to the footing may be serious and should be investigated carefully. This could be caused by uneven settlement of the building and could indicate the wall is separating — one part settling more on one side of the crack than the other. The crack extending to the footing could also indicate a failure of materials such as deteriorating block, decayed mortar, or weak concrete.

**FOUNDATION CRACKS**

• Vertical cracks

• Angled cracks

• Horizontal cracks

• Shrinkage cracks

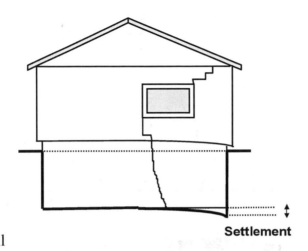

This diagram shows the results of settling due to soil weakness at one side of the house. Note that cracking on both the exterior as well as on the foundation wall are indications of the settlement. In this case, one part of the house is pulling away from the other.

**Settlement**

**Angled cracks** appear when the upload and download offset each other. They can appear when there is a major difference in the soil under the house from one location to another, heaving of the soil, and resulting footing failure. This type of crack points down to the location of the upload. In block construction, the angled crack may appear along the block joints in an angled direction. This is called a **step crack** and is shown here.

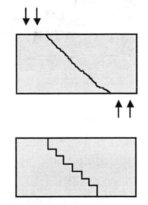

Angled cracks can appear in pairs, where a load in one direction is offset by a pair of loads from the other direction.

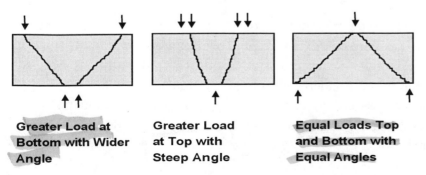

**Greater Load at Bottom with Wider Angle**

**Greater Load at Top with Steep Angle**

**Equal Loads Top and Bottom with Equal Angles**

## V-CRACKS

When a crack increases in width along its path, this is called a V-crack. A <u>vertical V-crack</u>, wider at the top, indicates the wall is settling at the ends of the wall more than in the middle. A <u>horizontal V-crack</u>, as seen in the side diagram of the wall on this page, indicates the wall is moving out from the middle more than from the top and bottom.

Watch out for V-cracks!

The house diagram shows the result of settlement, where the stress is born at one corner of the house. Here, the angled step crack is likely to have a companion crack on the adjacent wall at that corner, indicating that corner is breaking away from the rest of the house entirely and sinking. When a crack is at a single corner of the house, it can indicate a broken footing because of the condition of the soil underneath, expansive clay soils, or even the uplift from heavy tree roots in that location.

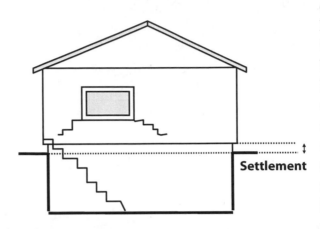

**Settlement**

A **horizontal crack** on the foundation wall can be an indication of pressure being applied from outside. The cause can be soil pressure against the wall, improper backfilling, and surface problems such as poor down-spouts that increase the amount of water pushing against the wall. The wall will bow inward and crack horizontally.

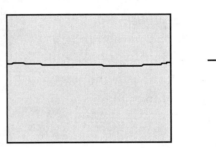 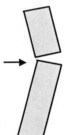

If the wall is actually **displaced** — when the surface of the wall is out of alignment — and the crack is wider on the inside face of the wall, serious problems can result. Concrete block and masonry walls will usually show the horizontal crack along the mortar joint.

There are other causes for horizontal cracks in foundation walls. There could be a vertical overload on the wall. Or the crack could be caused by settlement, where the wall has dropped when the soil underneath settled. In this case, the home inspector would probably not see any lateral displacement. Cracks may also be caused by heavy equipment during backfilling.

The home inspector may see **pilasters** — masonry or block columns against the interior foundation wall — built to support the wall and help absorb the lateral load. Pilasters may be part of the original construction or erected later as a corrective measure.

**Shrinkage cracks** can appear in foundation walls as part of the curing process. In the concrete block wall, you may see symmetrical step cracks that indicate block shrinkage. If the cracks appear around each block, this could be an indication of block shrinkage due to too much moisture in the blocks when they were set. It could also be the result of mortar that was placed in cold weather that later froze and expanded before curing. The home inspector may see a vertical shrinkage crack in the middle of a concrete block wall. The crack can be wider at the top indicating shrinkage during cure, where the bottom of the wall is held firm by the footings.

In the poured concrete foundation wall, shrinkage usually occurs naturally in the first few months. Some poured concrete walls are designed to have control areas, like seams, where cracking can be localized. If these areas are not provided, cracking is likely to be random over the surface of the wall.

## Using Your Judgment

Reading cracks as a clue to what is happening to the structure of a home requires judgment on the part of the home inspector. Cracks may indicate anything from serious structural problems to merely cosmetic or surface damage. The home inspector must learn when cracks are serious enough to suggest that customers monitor cracks or seek advice from a structural engineer. Here are some general guidelines to follow:

- Surface cracks that don't pierce the wall and show no displacement in any direction are not structural in nature.

- Cracks less than 1/4" wide, with or without displacement in any direction, are not likely to be signs of serious failure unless they are active.

- Cracks over 3/8" wide should be examined carefully as an indication of a potentially serious problem.

- Cracks that are still active should be noted as major defects. Active cracks will have sharp edges, will break a

*Definition*

*A pilaster is a masonry column built against a wall to help absorb the horizontal load and stiffen the wall.*

*Personal Note*

*"I urge you to realize how complicated the diagnosis of foundation cracks can be. Unless you're a structural engineer, you should constantly remind yourself that you don't know everything. And be extremely careful about suggesting solutions to customers. You'll probably only give uneducated advice once. Then, after the lawsuit, you'll learn to keep your mouth shut when you're not sure.*
*"What you do need to know is when to tell customers to monitor a crack and when to advise them to call in a structural engineer to assess the problem."*

*Roy Newcomer*

*Take a drive through several neighborhoods and pay attention to the general outline of the structure. See if you can spot sagging ridge lines, racking walls, chimneys pulling away, and settling portions of various homes.*

new painted surface or mortar repair. They indicate a failure in progress and a situation that can grow worse.

- Don't come to a conclusion based on a single crack. Investigate further. A crack in one area of the foundation will often have a corresponding crack in another area.

## Inspecting from the Outside

The home inspector begins the inspection of the foundation from the outside of the house. Let's go through the list of what to watch for on the exterior, one item at a time:

- **The ridge line of the roof:** The home inspector should sight the ridge line of the roof to make sure it is a nice straight line. Sags and distortions in the ridge line can be an indication of the failure of the structural members of the roof. But they can, along with other indications, be an indication of problems stemming from the foundation.

*In **Photo #10**, you'll see a **Lannonstone house**. Unfortunately our picture isn't quite good enough to show what was going on with this ridge line. However, to the naked eye, the ridge line was higher in the center and sloping off in both directions. Look at the slope in the first-floor windows. This house is sinking on both ends. The ridge was actually broken at the center, and the house was breaking in half.*

#10 Lannonstone house

- **Racking of the house:** Racking refers to a structure that is leaning in such a way that the angles it forms with the

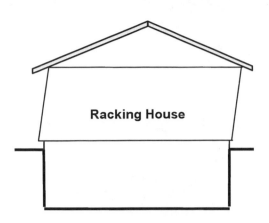

**Racking House**

foundation are no longer 90°. Racking generally indicates a failure of the wall structure, such as poorly braced walls. If you notice racking, keep it in mind as you examine cracking and other signs that something might be amiss with the foundation.

- **The chimney pulling away from the house:** Part of your inspection of the foundation involves checking out the foundation of the chimney. A chimney sits on its own foundation, which can experience the same settlement problems that the main foundation can have. Footings can fail due to soil weakness below.

**#11 Stone chimney pulling away**

*Photo #11 shows a stone chimney pulling away from the house itself. (Chimney inspection is not included in the structural inspection. It's considered to be part of the roof inspection. See A Practical Guide to Inspecting Roofs.)*

- **Twisted siding:** Framed houses with siding sometimes hide damage to the inner wall due to foundation settling. But often the damage can be seen in siding that is twisted or bent from the settling.

*Photo #12 shows **twisted siding** at the front of the house. Obviously, something is going on here that requires further investigation. This could be an indication of settlement of the foundation. As further investigation showed, this house happened to have a crawl space with such poor ventilation that the sill around the foundation had completely rotted out, and the house settled down to the top of the foundation wall. The foundation itself was in good condition.*

#12 Twisted Siding

#13 Another piece of twisted siding

*Photo #13 shows **another piece of twisted siding** — look on the second floor just along the lower roof. This needs further investigation. Another sign with this house is the fact that the stoop is settling. As a matter of fact, there are other signs of problems with this foundation, as you'll see in the next paragraph.*

- **Movement of foundation walls and cracks:** The home inspector should walk around the entire exterior of the house to check the foundation.

With **stone and brick masonry** foundations, the exterior and visible portion of the foundation should be checked for movement and cracks as well as the condition of the **mortar**. Use the tip of your screwdriver or a knife to probe the mortar. With brickwork that has water splashing against it from faulty downspouts, enough mortar can wash away to cause bricks to fall out. If water has been absorbed into the bricks, the face surface can **spall** or crumble away.

*Definition*

Spalling *is the crumbling and falling away of the surface of bricks, blocks, or concrete.*

#14 Block foundation

*Photo #14 shows the **block foundation** of the house with the twisted siding mentioned above. The wall is obviously cracked and displaced. In fact, it's pushing one block out from the corner of the foundation. This house appears to have serious foundation problems and needs to be inspected from the inside.*

#15 Another block foundation

*Photo #15 shows **another block foundation** from the exterior. Here, the block is cracked and displaced. Notice the jagged, sharp edges of the cracks, indicating an active crack. There is some surface deterioration on the block. What happened is that the wall leaned in and the pressure from the basement beam is forcing the block out.*

- **Corners of the building settling:**  Pay particular attention to the cracks in the foundation and exterior walls at the corners of the house.  Often times, footings will fail at the corners due to soil weakness or expansive soils.

- **Cracks in exterior walls:**  Check for crack patterns in the exterior walls.  Some cracks are cosmetic cracks or an indication of a problem with the wall itself, but others may indicate a settling problem.

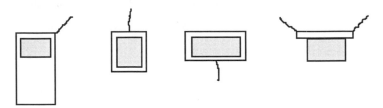

Cracks starting at windows and doors and ending a short distance away can be a result of overloading the header or lintel at the opening.  Here are some examples of cracks that are most likely cosmetic in nature, unless the cracks are large and active.

Other exterior wall cracks should be inspected carefully.  Photo #16,  is a good example of that.

*Photo #16 shows a **brick veneer house** with a step crack at the corner that turns into a vertical crack going down to the foundation.  This house had similar cracking on the adjacent wall — a sign that settling is occurring at this corner.  This should be investigated further.  When a house is brick veneer, you'll want to be sure that the problem is with the foundation and not just with the veneer itself.  Veneer can crack because it is not properly tied to the framework of the house.  In this case, it turns out this house was built on foundry sand and was indeed settling into the ground.*

#16 Brick veneer house

**#17 Stucco finish**

*Photo #17* shows a house that has a **stucco finish** with stones inlaid. This house had several large vertical cracks at various locations around the house. This is a serious situation. This house had been built on a landfill and the footings were sinking.

**#18 Step crack in a brick wall**

*Photo #18* shows a step crack in a **brick wall** that extends down to the foundation. A crack like this may not indicate a serious problem since the house is old and the crack appears to also be old. However, in this case, the crack has been patched and has reopened, and there appears to be displacement of the bricks. This needs to be investigated.

*Examine as many foundations up close as often as you can while you study this chapter. Bother your friends. Look at their foundations from the exterior and ask to see the basements.*

- **Displacement of windows:** Another sign to check for is the distortion of the window framing, which can also indicated settling of the house. Displacement of windows will often be accompanied by other signs of settlement such as exterior wall cracks.

In almost every case where some sign of footing settlement or foundation problem has been mentioned here, we've suggested to look further. This is true. Don't come to a conclusion without inspecting the foundation from the inside. The home inspector should point out cracks or other exterior signs to the customer while viewing them and explain that the situation will be investigated from the inside of the foundation.

One additional thing to watch when inspecting the foundation from the exterior is to make note of where the **land slopes away** from the house, exposing much of the foundation at the lower side. Obviously, the frost line is lower on the low side of the slope, and the footings should be stepped down in this area. Be sure to check this side of the foundation once you get inside.

*Photo #19 shows a **window in a brick house**. Notice how the bottom sill and frame are sloping down to the right. It's no longer in square with the rest of the window — an indication that the house is settling in that direction.*
*Note also the crack at the junction of these two walls. The two walls have pulled away from each other slightly. Again, it should be stressed that some settling happens in old homes over time. The situation may have already stabilized with no further problems expected.*

**#19 Window in a brick house**

## Inspecting from the Inside

The home inspector continues the inspection of the foundation from the inside of the basement or crawl space. It's important to keep in mind the findings from the exterior of the house and to search for answers for these signs.

- **Cracks in foundation walls:** The inspector should inspect *all* interior foundation walls, piers, and piles. First, let's talk about visibility and access — crawl spaces can be inaccessible and basement finishes and storage can hide portions of the foundation wall. Home sellers can be clever about stacking mattresses and boxes in front of walls they don't want the home inspector to see. The home inspector is not required to move any items that obstruct such access. However, it pays to look behind storage as much as possible to see if anything is hidden.

The home inspector must be careful to report what can be seen and what can't. We'll talk more about this starting on page 40, when we present how to report your findings.

#20 Poured concrete foundation

*Photo #20 shows a brand new poured concrete foundation with an angled crack. This crack ran down to the footing and met a vertical crack at the corner of the foundation (not seen in this photo). The wall was displaced. That is, the area above the crack was 3" further in than the rest of the wall. The wall was pushing against the gas pipe you can see at the top of the photo, creating a potentially dangerous safety hazard. This is a serious situation, a structural failure in progress, and we recommended the customer call in a structural engineer. Further investigation revealed that a backhoe had driven close to this wall and exerted enough pressure to pop the foundation. Notice also that there is evidence of water penetration at the footings, which is a signal of a drain tile problem. It turned out that this new home didn't even have drain tile.*

*Photo #21* *shows a fairly wide* *vertical crack in an old concrete block foundation.* *This crack is obviously more than 1/4" wide. Sometimes cracks that come from the window opening are not indications of foundation failure, especially if the crack does not extend down to the footings. And this one did not. Also, an old crack may show no signs of being active any more. This foundation showed no other problems. We concluded that the crack was old and inactive. Whatever situation had caused it was now stabilized. We recommended that the buyer monitor the crack for any new movement.*

#21 Verticle crack in an old concrete block foundation

#22 Horizontal crack in a concrete block foundation

*Photo #22* *shows a* *horizontal crack in a concrete block foundation.* *This is a fresh and active crack with sharp edges which are breaking the paint surface. It's at the frost line, indicating there's probably a problem with water pressure, grading, or faulty downspouts. (The frost line is usually about 3 or 4 blocks down from the top of the foundation.) We recommended that a structural engineer evaluate this situation.*

**#23 Step crack in a concrete block foundation**

**#24 Displacement of 4"**

*Photo #23 shows a **step crack in a concrete block foundation** where the top block is shifted over. There was considerable movement in this wall — a displacement of 4" as you can see in **Photo #24**. Although there were no signs that the patch job was not working, this situation was severe enough that we recommended an engineer come in for an evaluation.*

PLEASE NOTE: Most of the examples we're showing here are pretty bad and do require a structural engineer to evaluate them. We aren't suggesting that you should recommend an evaluation by an engineer *every time* you see a crack. Don't raise this alarm when you don't need to. You won't get any more referrals from real estate agents if they figure you're going to suggest an engineer for every house. We'll talk more about this later.

*A <u>bow</u> in a wall is a vertical curve, where the wall has an outward curve from top to bottom.*

*A <u>sweep</u> is a horizontal curve in a wall, where the wall has an outward curve from side to side.*

*A <u>bulge</u> in a wall is a combination of both bow and sweep.*

- **Bowing or leaning in foundation walls:** A foundation wall can have an inward or outward curve in it or lean inward. A **bow** is a vertical curve in a wall, a **sweep** is a horizontal curve, and a **bulge** is a combination of the two. They can be caused by the load from behind the wall such as a boulder in the backfill or from a narrow load from above. When a wall shows these distortions,

**Tilting But Still Stable**    **Tilting And Not Stable**

they may or may not have cracks associated with them. The danger is that the wall's center of gravity can move beyond its base, causing the wall to collapse.

The center of gravity is at the midpoint of the mass of an object. The object will overturn when the center of gravity moves outside its base. In a wall, if the center of gravity has moved more than 1/3 from the center of its base, the wall can collapse.

The home inspector can stand along a wall and sight down the wall to spot bows and leans. A level or a plumb bob can be used to measure any displacement in the wall. A flashlight shined along the wall during sighting can sometimes help the inspector to see bows in the wall.

*Photo #25 shows **bowing in a concrete block wall**. In this case, the bowing was causing some cracking as well. This pilaster was put in after the wall began moving. Notice that the pressure from the wall has caused the pilaster to open, indicating that wall movement is still active. We thought that the situation was still active and had not been fixed, and recommended that an engineer come in to evaluate the wall.*

**#25 Bowing concrete block wall**

- **Window displacement:** When inspecting a foundation wall from the interior, always take note of the basement windows and check them for displacement. This is a clue that the wall is moving.

- **Deterioration of materials:** The home inspector should identify the materials used in the foundation construction and inspect them for any decay or deterioration.

- **Water penetration:** The home inspector should inspect the foundation for water penetration into the interior. This is an important part of the foundation inspection. We're going to discuss inspecting for water penetration and drainage problems when we talk about inspecting other aspects of the basement, starting on page 44.

**#26 Total horror story**

*Photo #26 is a total horror story. There are so many problems here that you cringe to think about them. But take a look at the window. This photo is not distorting the angle at which this window is being pulled out of vertical alignment. It's really that bad. Notice also the bowing in the wall and the tremendous cracking that has occurred and been patched. The cracks are still active and are breaking through the patches in a vigorous manner. These signs were present throughout the basement. Obviously, a structural engineer was called in for evaluation. Even if this wall had been covered with paneling, there would have been enough clues visible to warrant an expert look at it.*

***Personal Note***

*"I once inspected a concrete block foundation with a steep step crack coming down from a window in the corner of the basement. There was some movement in the wall, and the paint was coming off the wall in sheets in that area. It looked very serious to me, so I recommended someone come in to evaluate the situation.*

*"I went back to the house with the engineer to see what was up. The house had been built on foundry sand and had settled about 12 years ago. But the engineer felt the crack had not opened for years and that all movement was done. He said it was nothing to worry about.*

*"It turns out the wall had been spray painted with some kind of coating that wasn't adhering to the wall — a paint problem having nothing to do with structure or water penetration.*

*"It was one of the worse basements I'd ever seen, yet the engineer gave it a clean bill of health. Of course, I'm not sorry that I recommended him. It's better to be sure."*

*Roy Newcomer*

*Photo #27* shows a **stained brick foundation**. *When you see something like this, be sure to probe the mortar joints and the bricks themselves for deterioration. Here, there's evidence of water penetration along the floor and staining that rises up the walls. The bricks have drawn up this moisture. Be sure to bend down and probe in the area of the staining. Mortar can deteriorate, becoming sandy, weak, and even wash out under these conditions. Bricks can deteriorate too. If the situation is serious, you can recommend that a mason be called in to look at it. Notice the doors in the corner. We looked behind them to see if anything was hidden there.*

#27 Stained brick foundation

*Photo #28* shows an **old limestone basement** *that on first glance looks in terrible shape. What you do see is a deterioration of the finish over the stones. However, the condition of the stone wall was okay. This photo is interesting because of a feature we've often seen in these types of walls. Notice the horizontal line on the far wall. This is not a horizontal crack. A 1 x 2 length of wood had been built into the wall. We probed the wood and found it to be totally rotted. On further probing, we discovered that the "crack" was only 2" deep and did not penetrate any further. We still don't know why the wood was there.*

#28 Old limestone basement

## Let's Review

Use your judgment. Know what you've found and what constitutes a foundation in good condition and one with defects. (Remember that we haven't yet talked about water penetration through the foundation yet. The comments below are regarding settlement and movement of the foundation and cracking of the foundation walls.)

- **When to give the foundation an okay:** You could find nothing at all wrong with the foundation, with no indication of movement, settlement, or wall cracks. There could be minor surface cracking that doesn't pierce the wall and with no displacement in any direction, which would not indicate a structural problem.

- **When to have the customer monitor the situation:** You may find cracks with displacement in the foundation wall that indicate that some settlement has taken place, but all signs point to the cracks no longer being active. For example, an old basement with an old paint job may have a crack, but there's no evidence that the crack has broken the paint seal. This indicates no more movement has taken place. But to be on the safe side, you should suggest that the crack be monitored to be sure, which the customer can do.

 There may also be cracks less than 1/4" wide that appear to be active, but are showing little or no movement. Suggest that these be monitored for now, but mention that a structural engineer should be consulted in the future if the cracks become wider and more displaced.

- **When to recommend a structural engineer:** The problem is potentially serious when you find active cracks over 3/8" wide indicating ongoing movement or settlement, horizontal cracks with movement, step cracks indicating footing failure, excessive bowing or leaning of the wall with or without cracks, and shearing in the foundation. These are indications of a structural failure in progress and should be investigated by an expert in these things. Recommend that a structural engineer be brought in now.

*Personal Note*

*"If you do recommend a structural engineer be called in to investigate a foundation, go back to the house when the engineer comes in. Find out what the engineer thinks of the situation. It's a great learning opportunity for you and will contribute to your own skills. It will also let you know if you're overreacting and raising a red flag when you don't need to."*

*Roy Newcomer*

A NOTE ABOUT MONITORING: First of all, we would suggest that the customer have the cracks patched with mortar and then monitor the walls for further movement. A wall crack can be monitored for further movement in several ways. The homeowner can make marks on each side of a crack, measure the distance with an accurate measuring device (steel ruler, screw adjustment dividers, etc.), and re-measure the crack every 3 to 6 months. Another method is to glue a microscope slide across the crack. If there is further movement, the glass slide will shatter or break away. Stiff paper or plastic can be glued across the crack. These materials will twist or deform if there is further movement.

## Reporting Your Findings

When you're inspecting the foundation, have your customer present. It's a smart practice. When the customer comes with you, you have an opportunity to fully explain the inspection and point out findings to the customer. Customer knowledge is a big step toward the prevention of complaint calls later.

Keep a running dialogue going with the customer. Not everyone is familiar with foundation construction and problems, and they may not understand what you're doing and what you're finding. Because some foundation problems can represent a major repair to the home, you want to be sure that the customer understands what you're saying. So **keep it simple**, but do talk about it. And pay attention to whether the customer understands. During the inspection, be sure to explain the following as you go along:

- **What you're inspecting** — the foundation.

- **What you're looking for** — cracks, deterioration of materials, evidence of movement and settlement, and so on.

- **What you're doing** — probing the mortar, measuring wall displacement, sighting the wall, and so on.

- **What you're finding** — footing failure, cracks indicating wall movement, and so on.

- **Suggestions about dealing with the findings** — monitoring a wall, calling in a structural engineer or mason, and so on. But with this caution — don't make uneducated guesses about how foundation repairs should be made.

## Filling in the Report

Every home inspector needs an inspection report. A **written report** is the work product of the home inspection, and every home inspector is expected to deliver one to the customer after the inspection. Inspection reports vary a great deal in the

industry, with each home inspection company developing its own version. Some are considered to be excellent, while others are not very good at all. A workable and easy to use inspection report is important for a home inspector in terms of being able to fill it in. Of greater importance is its thoroughness, accuracy, and helpfulness to the customer. We can't tell you what type of report to use, but let's hope it's a professional one.

The **Don't Ever Miss** list is a reminder of those specific findings you should be sure to include in your inspection report. We list these items after years of experience performing home inspections. Missing them can result in complaint calls and lawsuits later. Here is an overview of what to report on during the inspection of the foundation:

- **Type:** Identify the type of foundation walls present, such as poured concrete or concrete block.

- **Wall condition:** Take the time in the report to accurately report on the condition of the foundation walls. For your own protection, if there are areas not visible for inspection, it's best to note that in your report. This is a protective measure taken in case a complaint comes in later about something missed during the inspection.

  If cracks or movement of the walls are found, record the precise location where the condition is found and then clearly define what you've found. This helps a structural engineer who may be called in to evaluate the condition.

- **Recommending monitoring:** If your finding is that a crack condition should be monitored, be sure to mention that fact in your inspection report. See page 39 for those instances that call for this recommendation.

• **Recommending a structural engineer:** Be sure to indicate clearly when a problem is serious enough to call for an evaluation by a structural engineer. Again, see page 39 for examples of conditions calling for this recommendation. Any foundation condition requiring a structural engineer's evaluation should be classified as a **major repair** in your report. It's a good idea to report this major repair on the foundation page of your report and on a summary page at the back of the report.

*Report Available*

*The American Home Inspectors Training Institute offers both manual and computerized reports. These reports include an inspection agreement, complete reporting pages, and helpful customer information. If you're interested in purchasing the Home Inspection Report, please contact us at 1-800-441-9411*

# WORKSHEET

*Test yourself on the following questions.*
*Answers appear on page 44.*

1. A crack that increases along its path is called:

    A. An angled crack
    B. A horizontal crack
    C. A vertical crack
    D. A V-crack

2. Which is <u>most likely not</u> a cause of a horizontal crack?

    A. Broken footing at a corner
    B. Soil pressure
    C. Water pressure against the wall
    D. Poor downspouts

3. If you note an active 1/4" step crack with no apparent movement, you should:

    A. Recommend that you come back to repair the crack.
    B. Suggest the customer monitor the crack.
    C. Recommend a structural engineer be consulted to assess the crack.
    D. All of the above

4. In Photo #25, what is the structure in the middle of the photo?

    A. A shrinkage wall
    B. A pilaster
    C. A sweep

5. In a solid brick house, what is a sign of corner settlement?

    A. Step cracks at an exterior corner going down to the foundation
    B. A pair of angled cracks above a corner window
    C. Deteriorating mortar at the corner
    D. Spalling on the bricks at the corner

6. What is <u>not</u> a cause of foundation settlement?

    A. Soil weakness at the site
    B. Soil compacting underneath the foundation
    C. Soil pressure against the wall
    D. Footing failure in one or more locations

7. **Case study:** You find the west wall of a basement paneled and cannot inspect the foundation at that side. The north wall has a vertical crack under a window that doesn't reach the foundation. It's only 1/4" wide but seems to be active. Refer to Photo #22 to see what's happening to the south wall. At the middle of the east wall, there's a huge pile of storage you can't see behind.

    Which of the following would you record in your inspection report?

    A. Foundation wall condition satisfactory
    B. North foundation wall requires repair.
    C. West foundation wall not visible, unable to inspect.
    D. Storage in basement, did not inspect foundation walls.

8. For the case study above, for which condition should you recommend an evaluation by a structural engineer?

    A. Crack in south foundation wall
    B. Crack in north foundation wall
    C. Storage on east foundation wall
    D. Paneling on west foundation wall

9. For the case study above, would the condition of the north foundation wall be classified as a major repair?

    A. Yes
    B. No

# Chapter Six

# INSPECTING THE BASEMENT

The home inspector inspects the entire interior of the basement, including the foundation walls which we've already discussed. The basement inspection includes the following:

- Basement moisture and water penetration
- Basement drainage
- The basement stairs and/or hatchway
- Wall penetrations
- The basement floor
- The supporting structures — columns, sills, girders, joists
- Seismic bolts, where applicable

## Inspecting for Water Penetration

Leaking basements are a common problem. Generally, water problems in the basement are not dangerous to the home's occupants. However, water and excessive moisture in the basement are nasty and interfere with the use of the space. It's predicted that almost all, about 98%, of basements will leak at some time during their life, and can be categorized as occurring:

- During a **catastrophic event** such as a flood, hurricane, or an incident of broken plumbing.

- **Periodically** such as during the spring thaw or an unusually heavy rain when the ground water rises.

- **Constantly**, that is, with every rain fall or in an area with an abnormally high water table or where an underground spring is present.

Probably about 90% of leaking basements are caused by **surface water** — rain and snow collecting in the soil around the home and getting into the basement through the foundation walls and floor. Surface water should be made to flow away from the house and not be allowed to collect in the soil around the foundation. Constant leaking problems can be caused by:

A faulty gutter and downspout system
Improper grading of the land around the house
Nonfunctioning drain tiles due to silt buildup in the tiles
Patios, drives, and walkways tilted toward the house

The remaining 10% of leaky basements can be attributed to **ground water.** Basements can also have water or excessive

*Worksheet Answers (page 43)*

1. *D*
2. *A*
3. *B*
4. *B*
5. *A*
6. *C*
7. *C*
8. *A*
9. *B*

moisture in them due to problems other than actual leaking. Surface water leaking through the roof can make its way to the basement through the house. The plumbing system or appliances can leak, the sewer can back up, and there can be condensation.

When entering the basement, use your senses to get an overall opinion about moisture being present. Does the basement smell damp, moldy, or mildewy? Does the air feel damp and cold? Can you see other signs of leakage? Watch for these signs:

- **Standing water:** This is the most obvious sign, and the home inspector should investigate the source of this water.

- **Staining:** This is evidence of the previous presence of water in the basement. Stains on foundation walls and the basement floor can be old stains or recent stains, indicating when the problem occurred.

- **Efflorescence:** This is a whitish mineral deposit often seen on foundation walls. The crystals are left on the wall after water has seeped through the wall, bringing with it dissolved salts from the masonry or concrete. The crystals are left on the interior wall as the water evaporates.

- **Rust:** Even if stains have been cosmetically covered over, the home inspector can often see evidence of water problems by inspecting the base of steel supporting columns for rust. Other items that can rust are nails in baseboards or paneling, electrical outlets, or the metal feet on appliances.

- **Damage to the ceiling:** Water penetration into the basement from the house can usually be seen in the condition of the structural members in the floor above. The home inspector may find evidence of leaking in rotted subflooring and joists under bathrooms.

- **Damage to basement finishes and contents:** Keep an eye open for warped paneling, crumbling drywall, moldy carpeting or warped floor tiles, and soggy storage boxes or storage placed on palettes.

---

**MOISTURE INSPECTION**
- Standing water
- Staining
- Efflorescence
- Rust
- Ceiling damage
- Damage to finishings and contents

---

*Definition*

*Efflorescence is the white mineral deposit left after water passes through the foundation wall, having dissolved salts from the materials of the wall. It appears on the interior of the foundation after the water has evaporated.*

*"One of my inspectors was looking at a basement which had a counter-high cabinet built along one wall. The tiniest evidence of water stain appeared on the wall at the top of the cabinet. He opened its door to find water flowing out of a crack down the wall behind the cabinet. He could so easily have walked right by that cabinet without looking further.*

*"The point I'm making here is to keep your eyes open and to investigate. Look behind storage as much as possible. And open doors to cabinets and storage areas to see what's inside."*

Roy Newcomer

*Photo #29* *presents a* **basement with fresh water** *on the floor. What do you see here? Is the basement leaking or are the water heater and water conditioning equipment leaking? Or both? Upon inspection, we determined that the water heater and conditioning tanks weren't leaking. This basement had leakage of water from the outside along the junction of the wall and the floor. This is an indication of a* **drain tile problem** *— either an absence of drain tile or clogged drain tile. And there's evidence this is a fairly persistent problem. Note that the bottom row of concrete block is cracked, stained, and deteriorated. Leakage had been going on for some time. Note the storage racks built up off the floor, a clue that the owners have had an ongoing problem.*

It isn't always possible to determine what the situation with the wet basement is or what the source of the problem is. During the single visit to the property, it's difficult to get a fix on the severity of the problem and the frequency. For example, it isn't always possible to know if stains are present because of a one-time occurrence, an ongoing problem, or a problem that has since been remedied. But that doesn't get the inspector off the hook. Every effort should be made to figure out why the problem is occurring.

You might want to refer back to **Photo #20** to take another look at the poured concrete basement. Water penetration into this basement was also identified as a drain tile problem. If you remember, this is the basement that turned out not to have any drain tile at all. We caught this problem soon after the house was built, so you don't as yet see too much deterioration to the concrete at the base of the wall.

**#29 Basement with fresh water**

Sometimes water penetrates into the basement through cracks in the foundation wall itself.

Often you'll see patches on cracks with evidence of previous water staining on the wall or the floor below. If everything is dry at the time of the inspection, it isn't always possible to tell if the patch is working and the problem solved. The cracking occurred in the first place from excess water in the soil outside the foundation. So, unless the problem is corrected properly from the outside, interior patches will only cause the water to break through somewhere else.

#30 Standing water

*Photo #30* also shows **standing water** in the basement. Here there's evidence in the far wall of fresh water just at the base of the foundation wall indicating a problem with the drain tiles. Otherwise, there's so much water it's hard to tell exactly where it's coming from. The supporting columns are an interesting clue to how long this problem has been occurring. Notice that the left column is rusted out at the bottom, indicating persistent wetting. In fact, the column on the right appears to be a replacement or additional support provided for the rusted column.

*Photo #31* shows evidence of leaking in a **concrete block basement**, also a drain tile problem. There's no standing water, but there are damp patches showing recent leaking and water stains on the floor showing more extensive leaking in the past. This situation has been going on long enough and often enough for the block to crack and stain and for the mortar to weaken and fall away — and for owners to raise storage racks off the floor. The feathery white stains are efflorescence, from the minerals in the concrete being forced out by moisture in the walls.

**#31 Concrete block basement**

Van Lines

**#32 Severe cracking**

*Photo #32* shows a concrete block wall with **severe cracking**. The horizontal crack four blocks up from the bottom is letting water through the wall.

**#33 Water entering through the foundation wall**

*Photo #33 also shows an old basement with **water entering through the foundation wall**. Here there's plenty of evidence of water coming in from the top of the wall and through cracks. The mortar is washing away.*

*Although some of the white areas in the photo show attempts to patch this wall, the feathery white areas show efflorescence. When basement leakage is localized in a corner as shown here, check the downspout in that corner. This could be caused by a missing splash block or one that is out of position or poorly located. It can also mean that the landscaping is sloped toward the house.*

## Basement Drainage

The home inspector should note the presence or absence of a **floor drain** in the basement.

In some areas of the country, the floor drain will have a **palmer valve** in it, which you can see if you shine your flashlight down the drain. There's a little cover inside the drain on hinges that opens and closes to allow the water in the drain tile system to flow through there and go out through the sewer. This valve should be functioning. Sometimes, what appears to be a drain tile problem turns out to be a malfunction with the palmer valve, a much less expensive problem to fix. While quotes for drain tile repair or replacement can be upwards of $6000, it can cost only $35 for a plumber to break open a stuck palmer valve.

If your area makes use of the palmer valve and you've seen evidence of a drain tile problem, check out the floor drain to see if there's a valve. Suggest that your customer or the homeowner check with a plumber first to determine if the valve is causing the problem. You'll be a hero if a malfunctioning valve is the cause of the basement leakage and not faulty drain tile.

*Definition*

*A palmer valve is a hinged valve in the floor drain which allows water from the drain tile system to flow through the floor drain and out into the sewer. They are not used everywhere in the country.*

*For Your Information*

*Ask a plumber whether palmer valves are used in your area. Find out as much as you can about them.*

If there's a **sump pump** in the basement, the home inspector should check it to see if it's operating. As was explained earlier, a sump is a pit below the basement floor made of concrete or earth walls and floor. Water enters the sump from the perimeter drainage system through bleeders. An electric sump pump in the pit pumps water away from the house.

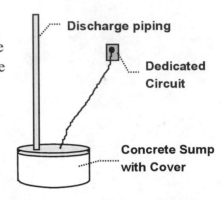

Discharge piping

Dedicated Circuit

Concrete Sump with Cover

The pump can be a **pedestal-style pump** which has the motor mounted on a shaft where it sits above the water level. A lever will stick out of the crock. To test the operation of the pump, pull up on this lever. The sump pump may be the **submersible** type, sitting below the water level. For this type of pump use a wooden stick to pull up the pressure switch or float in the crock. Another way to test each kind of sump pump is to run water into the crock with a hose (not required).

The sump pump motor should run quietly and should discharge water. It shouldn't run all the time. The pump should have its own dedicated electrical circuit so it continues working even if some other equipment malfunctions. The crock should be covered and kept free of silt buildup and debris.

## Stairways, Floors, and Penetrations

Part of the inspection of the basement is to inspect the entryways into the basement — interior stairway or exterior hatchway — the basement floor, and penetrations through the foundation wall such as venting and plumbing lines.

- **Stairways:** Interior stairways to the basement should be inspected for safety. The home inspector should check for proper lighting so people going down the stairs can see their way to the bottom. Handrails should be present and securely fastened. Steps should have risers of equal height and should be level, uncracked, unworn, and stable. The stairway should be examined at the points of contact with the basement floor. Sometimes, the feet of the stairway sit within the slab (when the stairway is built before the slab is poured) and may be in contact with damp soil. Check this area for wood rot or insect damage.

The home inspector will find the most interesting stairways in old houses. The stairs can be unreasonably steep and uneven, headroom can be insufficient, and handrails missing or loose. These conditions usually may not bother the customer who expects some quaintness in an old house, but these findings should be noted in the report nonetheless.

- **Exterior hatchways:** When there is an entry to the basement from the outside, the home inspector should check the covers (called cellar doors, bulkheads, or hatch covers) for deterioration and water penetration. The inspector should check the steps for safety and the condition of the masonry sidewalls.

- **The basement floor:** Although basement floors are usually not structural, the concrete slab plays a role in supporting the base of the foundation wall against lateral pressure of the soil. This is true with the wood foundation where the slab contributes to the support of the wall. The home inspector should note whether the floor is concrete or dirt and whether it is covered with carpeting or tile.

**MORE INSPECTION**

- Stairways and exterior hatchways
- Basement floor
- Penetrations through the foundation wall

#34 Stairway wall

*Photo #34 shows a common occurrence along the **stairway wall**. Notice where the above-grade wall meets the below-grade wall along the stairway? What's happened here is that the wood framing has experienced shrinkage and popped out the drywall at that point. This is not a structural problem.*

The concrete floor can have **shrinkage cracks** which occur during the cure of the concrete. It can also heave, sink, or crack in areas due to the action of the soil below. When footings fail, the home inspector may see **shear cracks** where the concrete slab is actually displaced. The diagram at the right shows the footing and foundation settling, causing a shear crack at the perimeter of the slab.

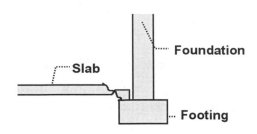

Slab · Foundation · Footing

- **Penetrations through the wall:** The home inspector should check all areas in the foundation wall, where there are penetrations to the outside, for water and/or soil coming in and for cracks around the area.

*Photo #35 shows **stress cracks** around an interior supporting column. This is evidence of the footing under the column settling. This is a potentially serious problem.*

#35 Stress cracks

The home inspector must watch for any evidence that the property has an old **underground oil tank**. In our area, it used to be acceptable for these old tanks to be filled in with gravel. But now they must be removed entirely. Evidence that the home used to have an oil burner can be seen if there are two penetrations in the wall. Either you'll see two holes where 1/2" or 3/4" copper tubing is sticking out or you'll see two patches over the holes. The tubing may be bent in or up. There could have been an *outside* or *inside* tank. If there are also two corresponding holes in the floor, that's where the tubing came in through the wall into the floor and up out of the floor to the old furnace. And that would mean there was or still is an underground tank. However, the new furnace was often placed over the floor holes.

You can't always tell if an underground oil tank exists, but the customer should be alerted to the possibility. The owner may have paperwork showing that the oil tank was filled in with gravel at some earlier date.

| UNDERGROUND OIL TANKS |
|---|
| Watch for evidence of an old underground oil tank and alert your customer to the possibility. |

***For Your Information***

*Find out what the situation is for old underground oil tanks in your area. Can they be filled in? Must they be removed?*

#36 Nonfunctioning openings

***Photo #36*** *shows* ***nonfunctioning openings*** *in the foundation wall. The opening on the right is an old septic line from the leach field. Staining on the wall shows where water has leaked in. In this case, there was also silt on the floor below the opening which indicates that soil is entering. A hole like this should be patched from the outside.*

**BASEMENT STRUCTURES**

- Columns
- Sills and headers
- Girders and joists

## Supporting Structures

From the basement, the home inspector will report on the condition of the supporting structures such as columns and the visible framing overhead.

- **Columns:** Posts and columns are vertical supports that carry the weight of the structure from the girders (or beams) to the ground. Columns transmit the weight to footings below. Often, you'll find columns or posts put in to support an unusually heavy load such as a piano.

  The home inspector identifies the materials used in column construction — steel, wood, or masonry. Columns should be inspected for their condition and ability to support the structure above. In **steel columns**, look for rust throughout the length of the column. Rusting at the bottom shows water present at the floor level. Rusting higher up may indicate water from above running down the column or an unusually high moisture content in the air or condensation. In **wood posts**, look for and report evidence of wood rot and possible insect damage. **Masonry columns** should be inspected for the condition of the mortar.

  The inspector should determine if the column or post is doing a proper job of supporting the structure above. First, look for footing problems (or the absence of footings altogether). Columns and posts should **sit on their own footings** and be **secured to the beam** overhead. They should be plumb. A column or post that is out of plumb by over 1/3 its thickness may no longer have its structural integrity.

  You may see a **shim** of metal or hardwood inserted between the beam and column. The shim should be large enough to cover the interface between the beam and column. If it's too small, the beam or top of the column may be crushed. Report on damaged or crushed shims.

  Refer back to **Photo #30** for a look at the two columns shown. The original steel column is rusted out at the bottom, and this should be reported. The column on the right looks like it's been introduced to provide the support lost by the first. The bottom on any new column should be inspected carefully. Is the column simply sitting on the floor? Does it have a footing of its own? In this case, the new column is free standing. This was reported.

Take another look at **Photo #35**. This column is in good condition, but the stress cracks and depression in the floor indicate that the footing below the column has settled. With a column like this, inspect the top of the column and see what's happening to its ability to support the girders above.

- **Sills and headers:** The **sill**, sometimes called the sill plate, or mud sill, is the portion of the framework that sits directly on the foundation and provides a pad for the bottom of the framing system. Sills today are 2 x 4's or 2 x 6's laid flat on the foundation and anchored in place with bolts. The sill in older construction may be an 8 x 8 wood beam. Wood sills support wood framing members, but not masonry, which sits directly on the foundation. The **header**, or rim joist, is nailed to the sill.

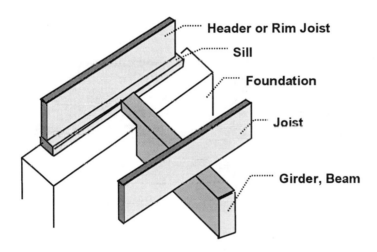

The home inspector should inspect the sill and header for rot. This can be caused by a defect in the siding, from soil too high around the exterior of the house, or water wicking up through the foundation. Suspicious wood should be probed with a screwdriver. Sometimes in cold climates, you'll find a pad under the sill or mortar packed against the sill. Check the mortar for condition. The home inspector may find insulation around the top of the basement wall, in which case the sill and headers may not be visible.

A sill over a window opening may sag or break from stress without a lintel or header having been installed underneath. This should be pointed out to the customer.

- Watch for deterioration, sags, cracks, and crushing.

- Be sure beams are secured to supporting columns and have proper end support.

- Check for improper notches, holes, and missing portions of the beam.

*Personal Note*

*"You simply won't believe the sorts of things I've run into when looking up into basement ceilings. Be especially careful where new heating runs or plumbing lines have been installed. Take the time to think it through. For example, they've cut through this beam. Does it still function? What does that do to the joists? What's supporting the wall above?*

*"Watch out too when you see repairs. Don't automatically accept the repair as successful. Think it through. A lot of handyman work is incredibly inept."*

*Roy Newcomer*

- **Girders:** Girders (or beams) are horizontal load-bearing members of a floor system that carry the weight of the floor and wall loads to the foundation and columns. Girders usually run parallel to the long side of the house, but there may be others, shorter ones running elsewhere. Types of girders are large wood timbers, steel I-beams, lengths of built-up lumber, or glue-lams. A new development is the pre-fabricated girder-like truss.

Girders rest in pockets of the foundation wall, shown in the diagram on page 55, on the sill, or on pilasters attached to the foundation. There should be 3" to 4" of girder resting on end supports and 1/2" of clear space around the wooden beam where it rests at the foundation. Wooden beams should also have a waterproof material such as metal or polyethylene film between the beams and the foundation.

- **Built-up beams** (a 3-piece 2 x 8 or a 3-piece 2 x 10) should have their butted ends staggered along the length and located over support columns and posts. The joint should be within 6" of the quarter point of the span of the beam. That is, if the beam is 12', the joint should occur 3' from either end, give or take 6" on either side.

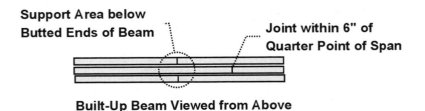

**Support Area below Butted Ends of Beam**     **Joint within 6" of Quarter Point of Span**

**Built-Up Beam Viewed from Above**

There are no simple rules to follow to tell if wood girders are properly **spanned,** center-to-center on their supports. It depends on the type and grade of the lumber. A 3-piece 2 x 10 beam should be able to span up to 10' in most 2-story homes; a 3-piece 2 x 8 can span up to 8' in a 1-story home.

There are, however, specific rules about what notches and holes can be cut into beams.

**Rule #1: Mid-notches** — where a notch is cut into a beam along its length — should be no deeper than 1/6 the depth of the beam. A mid-notch is allowable only

**1/6 of Depth**

at the top of the beam, not at its bottom. And the notch shouldn't be cut in the middle 1/3 of the beam's length.

**Rule #2: End notches** cut into beams — where the end of the beam rests on the foundation, sill, or pier — can be no more than 1/4 the depth of the beam. And as mentioned on the previous page, the beam resting on the foundation wall or pier should have 3" to 4" of beam resting on the end supports.

**1/4 of Depth**

**Rule #3: Holes** cut into beams should be no more than 1/3 the depth of the beam. Holes should not be cut into the top or bottom 2" of the beam.

**Hole 1/3 of Depth**
**Not within 2" of Edges**

**Rule #4:** Girder-like **trusses** should not be cut into at all or have any parts of the truss removed. The strength of the truss depends on the relationship of *all* of its parts to each other, and integrity can be lost by cutting or removing *any* part. Inspect trusses for loose or rusted fastenings.

Girders should be carefully inspected for *any* violation of these rules. Often heating contractors, plumbers, and electricians, who come in to work on finished houses, are the worst culprits. The home inspector may see evidence of their work.

**#37 Good Example**

---

**RULES**

- <u>Mid-notches</u> 1/6 of depth on top only and not in middle 1/3 of beam.

- <u>End notches</u> 1/4 of depth with 3" to 4" of overhang.

- <u>Holes</u> 1/3 of depth and not within 2" of edges.

- <u>Trusses</u> not cut or any parts removed.

---

*Photo #37 is a good example. Here, the heating contractor had cut an entire portion of the beam out! They added support posts and a brace at the right, but it doesn't appear to be a good solution. Notice the tilt. This situation was pointed out to the customer and recommendations were made to have this checked out.*

• **Joists:** Joists are horizontal members of a floor system that carry the weight of the floor to the foundation, girders, or load-bearing walls. Joists are supported by beams as shown on page 55. They meet the foundation wall, either resting on and nailed to the sill and header or resting on the foundation. Joists are generally wood — or more recently metal, plywood, waferboard, or wood trusses — 2 x 8's, 2 x 10's, or 2 x 12's, placed 12" to 24" apart.

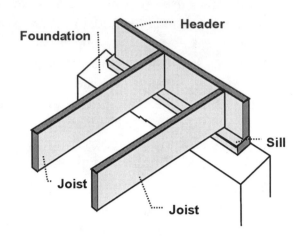

The allowable span for joists can vary considerably, depending on the material used. For example, a 2 x 8 joist of Douglas fir or yellow pine can safely span 12', but if made of spruce, redwood, or white fir only 10'. The inspector is usually unable to determine proper span without knowing the type of wood and is limited instead to determining the condition of the joist. Watch for deterioration and wood rot, cracking, twisting, sags, and loss of bearing.

Joists should be reinforced (doubled or tripled) under a partition. Other areas such as around stairway openings are reinforced with double or triple headers running perpendicular to the joist; for wide openings the joists themselves are doubled.

You'll often see a **bracing system** used between joists to add stiffness to the joists and keep them from twisting.

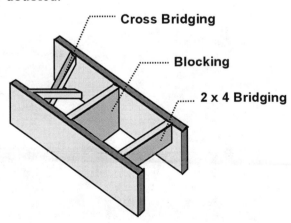

Building codes vary from area to area on how bracing may be done. You may see **blocking**, where the board nailed between joists is the same depth as the joists. Or **bridging** methods may be used in combination, where 1 x 3's are used diagonally in cross bridging and 2 x 4's are nailed between the joists.

The same rules that apply to girders regarding notches and holes apply also to joists. Joists can be seriously weakened by improper notching. Any violation should be reported.

**#38 Joists cracking**

**#39 Cracked joists**

*Definitions*

*Blocking* between joists is the use of a brace of wood the same depth as the joist which gives stiffness to the joists.

*Bridging* is a bracing method between joists where diagonal 1 x 3's (cross bridging) and/or perpendicular 2 x 4's are used to add stiffness to the joists.

*Take a look at **Photo #38** which shows **joists cracking**. **Photo #39** also shows evidence of **cracked joists**. Note that the joist at the upper left shows deterioration. It should be probed to determine the extent of deterioration. Note also the presence of a **mold-like material on the subflooring**. This house has some serious leaking problems from the bathroom above this area, causing damage to the subfloor and joists.*

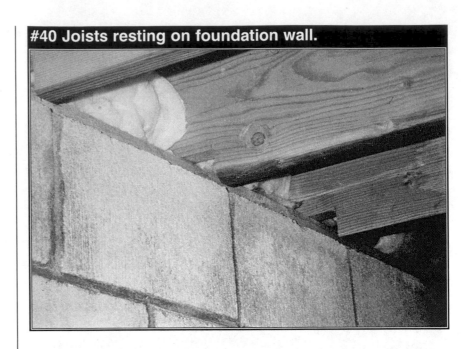

*Study **Photo #40** for a while to see if you can tell what's going on with these **joists** resting on the foundation wall. The ends of these joists have been notched to lower the floor. Notching like this will cause the joist to weaken and split. See the notching guidelines presented on page 56.*

- **Girder/joist connections:** There are several ways in which joists and girders can be joined. The joists can be **butt nailed**, **face nailed**, or **toe nailed** directly into the girder (not pictured here). In those cases, the home inspector should watch out for nails pulling out.

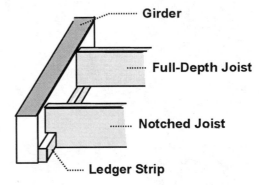

**Ledger strips** are smaller strips of wood nailed along the bottom edge of the beam. They may be supporting a notched joist or full-depth joist. These are weak connections between beam and joist. Watch out for nails pulling out. The ledger strip itself may be broken or sagging.

The home inspector may find **mortise and tenon connections** in older homes. The mortise is the hole or

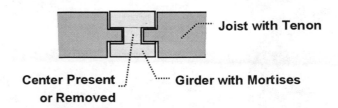

### Definitions

*Ledger strips are lengths of wood nailed along the bottom edge of a beam to provide support for joists.*

*A mortise is the hole or slot in a wooden beam that accepts the tenon, which is the projecting end of the joist.*

*A gusset plate is a piece of wood placed over partial bearing joists and nailed into the joists to hold them in place.*

slot in the girder; the tenon is the projecting end of the joist. The mortises can be back to back with the center portion of the girder remaining or with the entire center strip removed. This older method is weak, and you should look carefully for joist cracks at the tenons.

Joists can be **full bearing** on the girder or can be **partial bearing**. Full bearing can be done with or without overhang, as shown here. In the partial bearing method, joists can meet above the girder and be nailed in place with the use of the gusset plate or be notched or E-lapped over the girder.

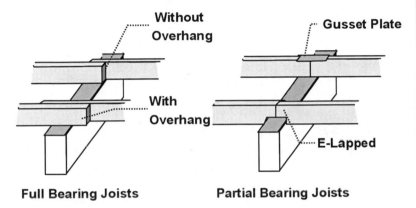

**Full Bearing Joists**       **Partial Bearing Joists**

Another method used in joist/girder connections is the **metal hanger** approach, also called joist hangers (not shown here). Some older construction will have metal straps or hangars supporting the joists. Newer methods are the use of sheet metal hangars, clips, or fasteners nailed to both girders and joists. These methods make for very strong connections, but are only as strong as the nails holding them in place. Any of these methods should be inspected carefully for loose or missing nails and for any damage to the metal parts.

## Reporting Your Findings

We've covered a lot of ground in this chapter. The basement inspection includes many items and is detailed. Talk to your customer while you're inspecting the basement. Be sure to explain what you were inspecting and what you found in each case. Take the time to answer questions. Customers may not understand the consequences of your findings and are counting on you to make sense of it for them. Be sure to show them where you've recorded the findings in the report. Suggest that they review the report again on their own after the inspection.

---

**JOIST/GIRDER CONNECTIONS**

- Butt, Toe, or face nailed
- Ledger strips
- Mortise and tenon
- Full or partial bearing
- Metal hangers or straps

---

**BASEMENT INSPECTION**

- Stairways
- Foundation wall cracks and penetrations
- The basement floor
- Girders and joists
- Water penetration, moisture, and drainage

---

| DON'T EVER MISS |
|---|
| • Old stains indicating water penetration |
| • Cracked floor joists |
| • Improperly cut or removed structural members |

Your inspection report should have a basement page in it with room to report on the items listed in the box at the left. For a start, be sure to record **access**, that is, whether you were able to gain access to the basement and to inspect it. Always make a note if your access was limited and let customers know that you can't find defects if you can't see them.

- **Basement moisture:** Be sure to indicate whether you've found moisture or evidence of past staining in the basement. Indicate whether there's standing water, fresh stains (must use a moisture meter) or old stains, and signs of leaking. And record where you found it — on the walls, floor, or ceiling, for example.

- **Floor:** Identify the type of flooring present. Be sure to indicate if the floor is covered with carpeting or tile and therefore not visible for inspection. Record if cracks were found.

- **Girders and columns:** Identify the materials used in girders and columns and report on their condition. Make a note of any that are stained or rusted, damaged, or in need of repair.

- **Joists:** Identify the type of joists present, whether trusses, wood, or metal. It's a good idea to record their size as well. Note any defects such as cracking, improper notching, and so on.

- **Sump pump:** For sumps, note whether or not one is present, whether or not it's been tested, and whether or not it's operating. Note any defects you've found. If the sump pump is not operating and needs replacement, you might want to include it in a **major replacement** category on a summary page of your report.

- **Stairs:** Report on the condition of the basement stairway. Be sure to stress safety factors such as uneven risers, missing handrails, and the lack of balusters if the situation calls for them.

| SUMP PUMP LIFETIME |
|---|
| Since sump pumps have a relatively short lifetime, it's a good idea to list them as items needing replacement within the next 5 years. Just for your protection. |

# WORKSHEET

*Test yourself on the following questions.*
*Answers appear on page 64.*

1. Water penetration in a single corner of the basement is <u>most likely</u> caused by:

   A. Efflorescence
   B. A faulty downspout or splash block
   C. Clogged drain tile
   D. A high water table

2. Which of the following does <u>not</u> indicate water penetration through the foundation?

   A. Water stains along the wall and floor junction
   B. Efflorescence on the wall
   C. Water stains below wall cracks
   D. Mildew in the subflooring

3. In Photo #31, which of the following indicate an ongoing water problem?

   A. Elevated storage
   B. Old water stains on floor
   C. Deteriorating bottom row of blocks
   D. All of the above

4. What if anything is indicated in Photo #35?

   A. The column is out of plumb.
   B. Water leaked in at this location.
   C. Stress cracks indicate a sinking footing.
   D. Nothing.

5. What is a sign that an underground oil tank may be present?

   A. A patched penetration hole in the wall
   B. Two holes in the wall with disconnected copper tubing
   C. An opening in the wall that's leaking
   D. All of the above

6. A notch cut into a girder or joist can be no more than:

   A. 1/6 of the depth
   B. 1/4 of the depth
   C. 1/3 of the depth
   D. 1/2 of the depth

7. A hole cut into a girder or joist can be located:

   A. Anywhere in the girder or joist
   B. Only within 6" of the quarter point
   C. At least 2" in from top or bottom

8. What is the weakest of the following girder/joist connections?

   A. Mortise and tenon
   B. Metal hangars
   C. Full bearing without overhang

9. Case Study: You find a basement essentially finished, complete with ceiling tiles, wall paneling, and carpeting. You notice that the paneling is warped along the bottom and the carpet has a white feathery stain along its edges. The paneling and the carpet are dry. Although there's a sump pump, the motor does not start when you pull up the lever.

   Which of the following comments would you record in your inspection report? *Circle as many answers as you wish.*

   A. Joists not visible, did not inspect.
   B. Effloresence on ceiling.
   C. Girders are steel.
   D. Carpeting should be replaced.
   E. Sump pump not operating.
   F. Evidence of moisture behind warped paneling, recommend further investigation.
   G. Floor is carpeted.

# Chapter Seven

# INSPECTING CRAWL SPACES

When it's possible, the home inspector must inspect the crawl space. We've already covered the inspection of **foundation walls** starting on page 22. And we've discussed how to inspect the basement for **leakage** and to inspect the **supporting structures** starting on page 54. These items are also part of the crawl space inspection plus the following items:

- The crawl space floor
- Ventilation
- Insulation
- Seismic bolts, where applicable

## Access

When the foundation is put in and the earth beneath the house is not removed, a crawl space is created. Building codes vary about the height of the crawl space. In some areas, it's as much as 36", but the height can be only 12" in others. Some crawl spaces are entirely inaccessible. Some standards or regulations provide guidelines for entering crawl spaces.

The more difficulty there is getting into a crawl space, the more likelihood there is that problems are brewing down there which have been ignored. The home inspector should make every effort to get into the crawl space and inspect it thoroughly. Yes, it can be a nasty job. Most standards state the following:

- *Required to enter underfloor crawl spaces except when entry could damage property, or when dangerous or adverse situations are suspected.*

- *Required to report the methods used to observe underfloor crawl spaces.*

Perhaps the most important point for the home inspector's protection against liability is reporting the **methods used** to access to the crawl space. Because serious defects may be present in the crawl space, the home inspector must let the customer know if the inspector was able to gain access to the crawl space to find those defects. The customer must understand that the inspector cannot be responsible for reporting these defects if access is impossible or limited.

*Worksheet Answers (page 63)*

1. B
2. D
3. D
4. C
5. B
6. A
7. C
8. A
9. A, E, F, G

## Inspecting the Floor

The floor of the crawl space may be dirt or gravel. One of the most common problems with crawl spaces is the deterioration of piping and the framing members above due to the moisture trapped in the crawl space. This is especially true if a dirt or gravel floor is left uncovered and if there is no proper venting of the space.

The dirt or gravel crawl space floor should be covered with a **vapor barrier** to prevent moisture from the soil being released into the crawl space. Vapor barriers, or retarders, can be polyethylene sheeting, roofing paper, blacktopping, or concrete.

**#41 Crawl space with a dirt floor**

*Photo #41 shows a **crawl space with a dirt floor**. In this case, the soil has not even been leveled. Note that the water heater is half buried, not a good idea. All of the piping you see in this photo will eventually rust out from the moisture that's being released from the soil. We told the customer this floor should be leveled and covered with concrete or a film vapor retarder. By the way, this is a good example of what you might find in a crawl space. It certainly illustrates the job cut out for the home inspector. We had to crawl around this space and look into the farthest corners. Areas such as these make the inspection of plumbing and heating equipment quite a chore.*

*Personal Note*

*"One of my inspectors once inspected a house that was 6 years old. The code inspectors had passed the house on all aspects. But the house happened to have a crawl space which my inspector inspected thoroughly. In the farthest side from the crawl space entrance, he found a long span of foundation wall sagging, indicating a serious footing failure. He suspected the entire wall would eventually fail because of it.*

*"This raised quite a stink with the code inspectors, but upon further investigation they discovered that someone had dug underneath the wall and no footing existed at all. The inspector was right.*

*"What's the point? Investigate the* entire *crawl space carefully."*

*Roy Newcomer*

*Obviously, it would be good for you to see crawl spaces for yourself. Get yourself a jumpsuit and locate some friends with crawl spaces. And go to it.*

## Ventilation

Perimeter vents are required in a crawl space to keep moisture from rising into the house. Ideally, there should be vents in all four walls, or at least two opposite walls, so there can be cross ventilation. The amount of venting depends on the size of the crawl space and whether it has a vapor barrier or not.

- With a vapor barrier, 1 square foot of free vent is needed for every 1,500 square feet of floor area.

- Without a vapor barrier, 1 square foot of free vent is needed for every 150 square feet of floor area.

The rules about when to keep vents open and when to close them are not exact. That's because of the differences in climate, location, the type of house construction, whether the crawl space is heated, and whether the crawl space is an extension of a basement. There is some agreement about closing vents in winter, but debate about whether they should be opened or closed in summer.

## Insulation

Mistakes are often made when insulation is installed in a crawl space. If the insulation is installed in the **ceiling** of the crawl space, the moisture deterrent layer should be located up against the floor above, with the batting facing down into the crawl space. The home inspector may find damaged insulation because of this mistake. When the insulation is installed upside down, the moisture from the crawl space passes around the vapor barrier and condenses in the batting, rotting it out and causing it to fall down. The home inspector should suggest that such insulation be replaced — this time with the moisture barrier *up*, facing the warm side as it should.

In the unheated crawl space, ducts and piping should be insulated unless wall insulation is installed.

## Structural Concerns

We're not going to review everything previously discussed on sills, columns, girders, and joists. But they are definitely part of the inspection of the crawl space and should be given the same attention as described in the basement inspection starting on page 54. Materials should be probed for deterioration. Girders and joists should be examined for condition and structural integrity. Nothing should be overlooked.

**#42 Crawl space with a dirt floor**

*Photo #42 shows another **crawl space with a dirt floor**. Look at this column. It was in good condition and had the proper footing. However, the soil was eroding away from the footing and seriously undermining the footing's ability to support the beam overhead.*

**#43 Crawl space handyman repair job**

*Photo #43 shows a crawl space **handyman repair job**. What happened here was the joists shrunk and slipped away from the supporting foundation wall.*

The solution was to insert a beam, supported by jacks, under the joists. What do you do when you see a situation like this? Although steel columns or wood posts with proper footings would have been the best solution to supporting the new beam, jacks can be acceptable under certain conditions. First, examine the jacks. The jacks must be the proper kind of **screw jack** used for these purposes. Under no circumstances should car jacks be used. Next, determine if they're attached to the beam. They should be securely attached. Next, determine if footings have been put in. Here, you can see a concrete footing is visible, but we dug around these footings to see if they were the right depth. Sometimes, you'll find that only an inch or two of concrete has been poured, and that's not enough. One more thing — the far jack in the photo has a shim inserted between the jack and the beam. This needs to be looked at. Determine if the shim was inserted because the weight of the beam has pushed the jack down, causing it to sink. If that's the case, this is not a good solution.

Photo #44 is an extreme example of what happens when moisture is allowed to enter the crawl space and cannot escape

#44 Gross deterioration of the flooring

*Photo #44 shows **gross deterioration of the flooring** above the crawl space.*

through proper venting. Extensive rotting can take place. Refer back, if you will, to **Photo #12** in this guide with a view of a house with twisted siding. If you recall, this house had a sill that rotted completely out, and the house had fallen to the foundation. This house was an absolute nightmare. Not only had the sill rotted away, but many of the structural members were in terrible shape. **Photo #44** shows the subflooring in one corner of this same house. The floor was totally rotted out here. The owners of the house had built a window seat upstairs in this area to cover the fact that the floor was shot. In fact, the owners tried to deny that a crawl space even existed under the house!

## Reporting Your Findings

We should emphasize again that the home inspector *must* investigate the crawl space if it's possible to get in and move around in there. Of course, the customer is not going to get into the crawl space with you. So when you come out of the crawl space, communicate your findings with that customer. Talk over what you were looking for and explain what you found. Take pictures of problems, if possible.

NOTE: Sometimes, a house has a partial basement and a crawl space. Each should have its own inspection, and each should be reported in your inspection report.

Here's an overview on what to report on the crawl space inspection:

- **Access:** Explain whether your access to the crawl space was complete, partial, or not at all. If some areas were not visible to you during the inspection, make a note of that.

- **Foundation condition:** As stated on page 41 for foundation inspection, report the type of foundation walls present and note cracks, movements, settlement, and water penetration found. If the condition warrants, be sure to write your recommendation to either monitor cracks or call in a structural engineer for an evaluation.

- **Floor:** Identify the type of floor in the crawl space as concrete, gravel, or dirt, for example. Note if a vapor barrier is present. You may want to recommend that a vapor barrier be added if a gravel or dirt floor is uncovered. Report on any evidence you've found on water seepage.

- **Structures:** As described on page 62 for basement inspection, write your findings on the materials used for girders, joists, and columns and report on their condition. Report on any signs of deterioration in these framing members and in the subfloor from excessive moisture in the crawl space.

- **Insulation and ventilation:** Note if pipes are properly insulated in the unheated crawl space and whether ceiling insulation is properly installed. For ventilation, note the presence or absence of vents.

We recommend that you spend a little time with the customer after the inspection reviewing the inspection report. Especially with the crawl space, which the customer didn't see, it's important to stress any findings you may have. Show the customer the page on which you've written your findings. Often, customers will forget what you said during the inspection, and this short review can remind them.

---

**DON'T EVER MISS**

- Cracks in the foundation wall
- Unstable bowing or leaning of the foundation wall
- Shearing action in the foundation
- Signs of water penetration
- Cracked floor joists
- Deterioration or improper cuts in structural members

---

*Pages 70 and 71 present a
discussion on inspecting slab
construction. You may want to
go back to pages 19 and 20 of
this guide to review slab on
grade construction techniques
before reading this short
chapter.*

# Chapter Eight

## INSPECTING SLABS

Slab on grade construction can be completely invisible to the home inspector. The outside edge of the slab is often finished with stucco, brick, or some other covering. And the surface of the slab is normally covered with subflooring and finish flooring.

First, the home inspector should make sure that the construction method *is* slab on grade. What appears to be a slab on grade may be a concrete floor resting on piers and beams or simply grade beams. Ask the homeowner if there is some doubt on this point.

Even though the slab is not visible, there are signs that the home inspector can look for when determining the condition of the slab.

- **Settlement:** The home inspector should try to determine if the slab has settled. Just as with any other foundation, soil can be washed away beneath the slab. There can be problems at the corners with downspouts and splash blocks, causing the slab to settle unevenly. Heaving soil under the slab can cause the slab to tilt or sink, even twist and skew. Settlement can be caused by a leak in the plumbing or heating system embedded in the slab, which undermines the soil below.

  Watch for **cracks** in the exterior foundation (stem) wall and the slab itself. Even exterior cracks may not be a reliable sign, since they may only be in the stucco finish. But noticeable cracks on the interior floor can indicate settlement. Cracks in the slab are not likely to be visible. But even when the floor is covered, cracks can sometimes be felt under foot or seen when a flashlight is used to side light the floor.

- **Shifting:** Slabs can shrink and pull away from the outer wall if not properly attached during construction. The slab can crack along the wall perimeter. As a result, shreds of slab may be left on the wall where the slab has become detached. The home inspector may find cracking along the floor edge.

- **Moisture:** The home inspector should report evidence of water penetration through the slab. Moisture can come up into the house through cracks in the slab or through the openings in the slab. During the inspection, the home inspector should pay attention to these openings, such as plumbing lines coming up under the sink. These openings are places where moisture and insects can come up from the soil into the home. Moisture along the edge of the slab can indicate the slab has broken away from the foundation. When the slab and foundation sinks and settles below grade, water penetration can be a constant problem.

- **Post-Tension Cables:** Post-Tension Cable ends can rust if exposed to the weather. Verify that the concrete plugs over the cable ends are intact. If the cable ends are visible or rust is present be sure to note it in your inspection report.

## Reporting Your Findings

Customers should understand that visual inspection of the slab is not possible. Explain that you're looking for *evidence* that something may be wrong with the slab, but that situations could be developing that are not yet apparent.

- **Not visible:** When you're reporting on slab inspection, it's a good idea to be able to check off a not-visible box every time.

- **Slab condition:** Note if you've found signs of settlement, cracks in the slab, and the presence of moisture or water penetration.

---

**DON'T EVER MISS**
- Cracked, shifting, or settling slabs
- Moisture penetration

---

# WORKSHEET

*Test yourself on the following questions.*
*Answers appear on page 74.*

1. Which statement is <u>false</u>?

   A. The home inspector is required to enter the crawl space when possible.
   B. The foundation walls of the crawl space do not have to be inspected.
   C. The supporting structural members visible in the crawl space should be inspected.
   D. The home inspector should report the method of access used to observe the crawl space.

2. Ceiling insulation in a crawl space should be installed:

   A. With the moisture deterrent layer down.
   B. With the moisture deterrent layer up.
   C. Either way.

3. Perimeter venting is required in a crawl space to prevent:

   A. Heat accumulation.
   B. Insect penetration into the crawl space.
   C. The air from becoming stale.
   D. Moisture from rising into the house.

4. In Photo #41, what may happen because of the floor condition?

   A. Foundation cracking and settlement
   B. Footing failures
   C. Rusted piping and rotting in structural members
   D. All of the above

5. What is <u>not</u> a likely cause of slab settlement?

   A. Shrinkage of the concrete slab
   B. Faulty downspout and splash block
   C. Soil washed from beneath the slab
   D. Leaking heating system in the slab

6. What can be a sign of slab detachment?

   A. Moisture under the sink
   B. Cracks in the foundation
   C. Cracks along the <u>slab edge</u>

7. **Case Study:** You enter a crawl space from an entrance at the south end of the house to find a very uneven dirt floor. The entire north half of the crawl space is piled so high with dirt, you cannot inspect it. You find step cracks in the adjacent walls at the southeast corner. The cracks are displaced over 1/2". There is a vent visible in the south wall.

   What comment would you make in the inspection report about the north half of the crawl space?

   A. Area not visible, did not inspect.
   B. No vent present in this area.

8. For the case study above, what comment would you write in the inspection report?

   A. Recommend evaluation by engineer
   B. No unusual cracks or movement apparent

9. In the case study above, what suggestions should be made to the customer?

   A. Level the floor.
   B. Cover the floor with a vapor barrier.
   C. Check for operating vents in the areas currently not visible.
   D. Check the downspout and splash block in the southeast corner.
   E. Have someone look at those cracks.
   F. All of the above

10. What condition could be listed as a major repair in the inspection report?

    A. The ventilation situation
    B. The step cracks

# Chapter Nine

# ABOVE-GRADE CONSTRUCTION

The home inspector should be familiar with various types of framing and construction methods. This and the following chapters will give you an overview.

## Wood Framing

The most common wood framing construction method used within the last 50 to 75 years is the **platform framing** method. With platform framing, one story is constructed at a time, using 1-story-high studs.

First, the first story is constructed. Subflooring is nailed in place over the floor joists. At the outer edge and at partitions, a **sole plate** is nailed to the header and floor joists. Studs are toe-nailed or end-nailed to the sole plate. Double studs are used at openings in the walls; multiple studs are used at corners. A double **top plate** is end-nailed to the wall studs. Then the process begins again. Joists are laid for the second story, with an outer header or rim joist, followed by the subflooring, the sole plate, studs, and the top plate.

*Guide Note*

*Pages 73 to 81 present information on wall construction. Interior framing is presented in Chapter Ten; roof structures will be covered in Chapter Eleven.*

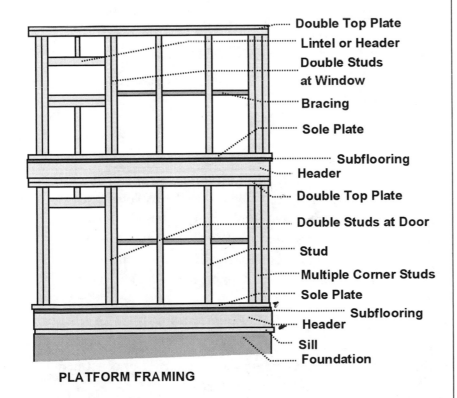

- Double Top Plate
- Lintel or Header
- Double Studs at Window
- Bracing
- Sole Plate
- Subflooring
- Header
- Double Top Plate
- Double Studs at Door
- Stud
- Multiple Corner Studs
- Sole Plate
- Subflooring
- Header
- Sill
- Foundation

**PLATFORM FRAMING**

## Definitions

*In platform framing, the stories of the house are constructed one on top of each other. The story-high wall studs are vertical framing members connected at the bottom to the horizontal sole plate and at the top to the horizontal top plate. Girts are horizontal bracing members used between adjacent studs as blocking. The lintel, a horizontal framing member, carries the load above a window or door opening.*

*In balloon framing, long vertical studs and corner posts run from the foundation to the roof, and the floors are hung on the wall frame. A horizontal ledger attached to the wall studs supports the second-story joists.*

*Worksheet Answers (page 72)*

1. *B*
2. *B*
3. *D*
4. *C*
5. *A*
6. *C*
7. *A*
8. *A*
9. *F*
10. *B*

It was the norm to use 2 x 4 studs, placed at 16" intervals, although today 2 x 6's have become common in energy efficient homes because they provide more space for insulation. The wood frame walls are load-bearing walls that carry the weight of the roof and floors down to the foundation. Metal studs are sometimes used for interior wall framing.

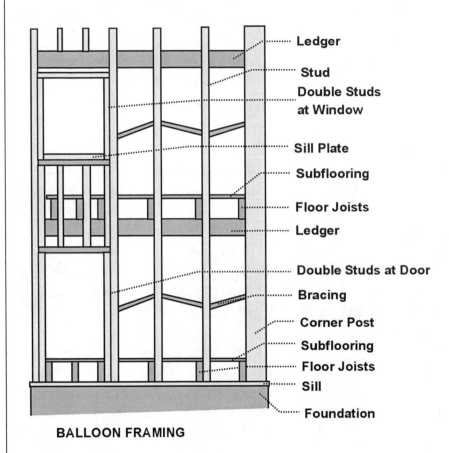

**BALLOON FRAMING**

Another construction technique home inspectors will run into was commonly used in the late 1800's and early 1900's. It's called **balloon framing**. This method used very long, uncut vertical studs and corner posts that ran from foundation to roof. Here, the studs and posts were erected first, sitting directly on the sill. The joists and flooring basically hung on the wall frame. A **ledger** was added for support of the second-story joists.

Both platform and balloon framing are equally stable when properly constructed. Platform framing replaced balloon framing in the early 20th century due to the ease of construction and the savings in terms of materials cost.

Of course, the home inspector won't see the wood framing of a home. It will be covered on the exterior and the interior. But there can be problems with the framing.

- **Faulty construction:** It's possible that a home can be constructed with overspaced studs. Door and window openings may not be adequately framed, and the inspector may be able to spot sagging above large windows if not bridged with the appropriate lintel. Nailing can be insufficient. Wall studs can buckle from overloading, especially if they're not properly spaced or lack blocking to brace them. Often times, you'll see framing that was originally built to bear loads, but a second- or third-story addition was added which increased the load beyond the studs' capacity to support it.

- **Inadequate lumber:** Poor quality lumber used in some wood framing can result in warped or bowed studs. Lumber that is too wet can shrink excessively after construction, resulting in warping and bowing. Both of these situations will probably be evident in the condition of the interior walls.

- **Deterioration of wood:** Wood framing can rot if there is water penetration into the structure and can suffer insect damage. Condensation can also occur when insulation is upgraded and vapor barriers are not provided on the warm side of the wall. During the cold months, warm moist air entering the wall from the house cools in the insulation and condenses, letting off water in the walls. Patches of peeling exterior paint can be a sign that condensation is going on in the walls.

<table>
<tr><td>**CONSTRUCTION**</td></tr>
<tr><td>• Platform framing</td></tr>
<tr><td>• Balloon framing</td></tr>
<tr><td>• Logs</td></tr>
<tr><td>• Post and beam</td></tr>
<tr><td>• Wood with brick veneer</td></tr>
<tr><td>• Solid masonry</td></tr>
</table>

## Post and Beam

Another type of wood framing construction is **post and beam**. Some prefabricated kits are available today using this type of construction method, but by and large, post and beam construction isn't commonly used in homes. You're more likely to see it in barns and large buildings such as churches and old mills.

The post and beam method uses framing members that are larger and fewer in number than the studs used in platform or balloon framing. A combination of posts and beams carry the weight of the structure down to the foundation.

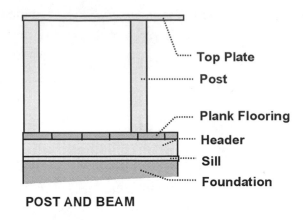

**POST AND BEAM**

In post and beam framing, 2" thick floor planks are typically laid. Some older floor planks can be up to 12" wide. The walls could simply be heavy plank, installed either horizontally or vertically and may not be load-bearing. A stud framework can be introduced between the posts to form the framework for a finished wall. In post and beam construction, fewer roof rafters were used, one at the top of each post, and planks used to form the surface of the roof.

These types of structures usually have a variety of connecting methods between the posts and beams — mortise and tenon, dovetail joints — and may not have used nails at all.

There are problems associated with post and beam construction. There can be a lack of rigidity to the structure when the exterior wall finish doesn't provide a stabilizing function. Because the load of the structure is concentrated at the posts, the foundation may be weak at these points. And the large timbers used in the structure may shrink and expand, causing the whole structure to shift with each changing season.

## Log Homes

Log construction is not always visible. In the old days, people often covered over a log home with stucco or clap board on the outside and finished the interior walls too. If that's the case, it's hard for the home inspector to even know about the logs or to assess any deterioration of the wall.

In traditional log homes, **chinking** — a mortar made of clay, sand, and other binders such as animal hair — was used to fill the gaps between the logs. Chinking needs annual maintenance and some parts will have to be redone each year. Also in the

*Definition*

*Chinking is a mortar made of clay, sand, and other binders such as animal hair that is used to fill the gaps between logs in a log home.*

traditional log home, the logs will shrink and expand across the grain of the log. That is, a log wall will get shorter and taller, causing gaps around windows and doors and pulling the floor up and down with it.

Modern log construction makes use of tooled logs with insulation between the faces. These logs are usually well dried before construction to minimize shrinkage. A plastic sealant is used instead of chinking.

## Brick Veneer over Wood Framing

Brick houses built today, and indeed since the early 1970's, are generally brick veneer over wood framing. A brick veneer wall does not carry the load of the structure.

The brick veneer is constructed from the foundation up and is attached to the wall sheathing with **brick ties**. These ties are usually crimped, accordion-style, to allow them to expand and contract

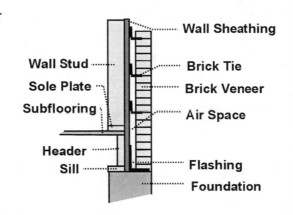

with the wooden frame and keep them from cracking the brick veneer.

An air space of about 1" is left behind the brick veneer to allow water passing through the brick to run down the wall. This water exits through the bottom row of brick through **weep holes** (not shown here) which are openings every foot or so along the bricks. A **flashing** runs beneath the brick and up the wall to prevent this water from reaching the foundation.

With brick veneer, the inspector is able to see signs of problems. Weep holes can become blocked and not allow water to exit from behind the bricks, leading to deterioration of the brick and the mortar. Brick ties can be improperly installed or loosened over time, and the veneer can separate from the wall.

*Guide Note*

*Another one of our guides (A Practical Guide to Inspecting Exteriors) will present more information about the inspection of brick veneer and other exterior sidings. This chapter is to be used to learn about construction techniques more than about the actual inspection of the exterior walls.*

**#45 Brick veneer house**

*Photo #45 shows a **brick veneer house** that had a separation problem. At the exterior windows, as shown in this photo, you could sight along the side of the window and see more brick showing at the top than at the bottom. Notice the cracking in the paint surface at the side of the window and along the top. The wall also has a bow to it. The veneer is pulling away.*

**#46 More brick**

*Photo #46 shows **more brick** at the top of the door than at the bottom. Notice also the small inset in the wall to the left of the door where the slant of the brick veneer is quite noticeable. Other sides of the house show extensive cracking of the brick veneer in the upper story, which helped us to determine the veneer at the first story was loose and separating from the house. A specialist needs to evaluate this situation.*

## Solid Masonry Walls

The walls of a home can be made of solid masonry. Usually, there are two thicknesses of masonry. The interior wall may be left exposed, plastered over, or covered with drywall. The masonry wall can be made of such materials as brick, stone, and concrete block.

A **solid brick wall** can usually be identified by the **header rows**, where the brick is turned with its small end facing out. The header rows serve as ties to hold the bricks together. (Header rows are not present in brick veneer walls.) The thickness of brick walls has declined over the centuries from 20" thick to 16" to 12" and finally to 8" thick. Most brick homes built since the early 1970's are, in fact, brick veneer over a wood framework.

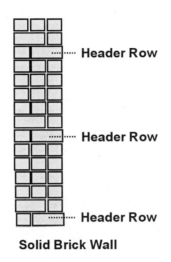

<sup></sup>······· **Header Row**

······· **Header Row**

······· **Header Row**

**Solid Brick Wall**

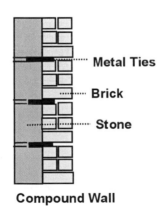

······· **Metal Ties**

······· **Brick**

······· **Stone**

**Compound Wall**

The solid masonry wall may be built of two materials — an inner and an outer layer. This is called a **compound wall**. Metal ties are used to attach the two layers together and give strength to the wall. The positioning of header rows in the brick outer layer can be random, not necessarily every five to seven rows.

If a dead air space is left between layers, the wall is called a cavity wall.

A masonry wall with an air space left between the inner and outer layers of the wall is called a **cavity wall**. In older construction, both inner and outer layers were often brick. A header row of bricks laid with the small end out, traversed the cavity and held the wall together. In more modern construction, the inner layer could be stone or concrete block. In the **compound cavity wall**, where you find different inner and outer materials, the brick could be attached to the inner layer with metal ties or with header rows.

### Definitions

In a _brick veneer_ house, an outer layer of bricks is attached to the wood framework of the house using _brick ties_, which are accordion-style metal fasteners. _Weep holes_ are openings in the bottom row of brick providing an exit for water accumulating behind the brick veneer. A _flashing_ placed at the foundation is sheet metal which prevents water leaks into the foundation.

In a _solid brick house_, three layers or wythes of brick are used to construct a solid wall with no wood framing. _Header rows_ are rows of bricks turned small end out to act as ties to hold the wall together.

A _compound wall_ is a solid masonry wall built of two different materials. A _cavity wall_ is a masonry wall with a dead air space left between layers.

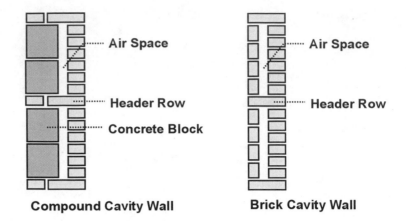

Compound Cavity Wall          Brick Cavity Wall

*For Your Library*

*If masonry walls are common in your area and you can expect to be inspecting many of them, it's a good idea to invest in some books on the subject. Look for titles that will educate you further in problems with masonry walls and appropriate repair work.*

The home inspector will note the following problems that may be seen with solid masonry walls.

• **Cracking:** Cracks can appear in a masonry wall for a variety of reasons. Some cracks are the result of foundation settling and movement. Refer back, if you wish, to pages 22 to 25 and pages 33 and 35 in this guide for the discussion on cracking involving the foundation. Refer also to **Photos #18 and #19**, discussed in these same pages, to see examples of cracking that appears as a result of settlement.

Other types of cracks appear because of some problem with the above-grade construction of the home. **Step cracks** above doors and windows indicate a problem with the lintel. A rusting lintel, which expands as it rusts, can lift the masonry above it, causing step cracks to move upward from the top of the window. A sagging lintel may cause step cracks in an inverted V-shape over the top of the opening or a **horizontal crack** at the top. There can be lower horizontal cracks and sagging above an opening in the foundation such as a hatchway.

• **Deterioration of brick and mortar:** Both brick and mortar can deteriorate if the right materials were not used in the original construction. It's possible that mortar could have been mixed with too much sand making it weak or with too little making it brittle. And bricks can have a surface that is too soft to keep out water.

But generally, **water** is the enemy of brick and mortar. Depending on how the mortar was finished off, rain may be allowed to enter into the wall and into the interior of the brick. Deteriorating mortar will wash away. The surface

face of deteriorating brick will crumble and fall away if water is absorbed into the bricks. This is called **spalling**. Deterioration may be present at the bottom of the wall where rain has been allowed to splash against the masonry. Both brick and mortar can be damaged from condensation inside the wall.

- **Bowing or leaning:** The masonry wall should be plumb, but there can be distortions such as **bow** (vertical outward curve), **sweep** (horizontal outward curve), **bulging** which is a combination of bow and sweep, and leaning for a number of reasons. Common causes are the deterioration of the mortar and the expansion and contraction of the wall itself.

  Masonry walls can be affected by the **interior framing**. Expanding or warping joists can cause the wall to bulge outward and even crack. Spreading roof rafters can push out the tops of walls. And, of course, foundation problems and movement can cause the masonry wall to distort.

REPORTING: When reporting on wall structure in your inspection report, be sure to identify the walls as wood framing or masonry. If you can't determine the wall structure in the home, make a note that interior wall structure was not visible. (The inspection and reporting of findings for exterior walls is presented in more detail in *A Practial Guide to Inspecting Exteriors.*)

# WORKSHEET

*Test yourself on the following questions.*
*Answers appear on page 84.*

1. Match the type of construction with the illustration.

   A. Balloon framing  *3*
   B. Brick veneer over wood framing  *2*
   C. Brick cavity wall  *4  8*
   D. Compound cavity wall  *8*
   E. Compound wall  *5  4*
   F. Platform framing  *4*
   G. Post and beam framing  *6*
   H. Solid brick wall  *7*

**Figure #1**

? 

**Figure #2**

*B* ?

**Figure #3**

*A* ? *Ledger*

**Figure #4**

*C*

**Figure #5**

*Tie In* ?

**Figure #6**  *G*

**Figure #7**  *A*

**Figure #8**

*D* ?

*2* 1. In Figure #1, what is the item marked?   *Bracing*
*3* 2. In Figure #2, what is the item marked?   *Flashing*
*4* 3. In Figure #3, what is the item marked?   *Ledger*
*5* 4. In Figure #5, what is the item marked?   *Metal Brick Tie*
*6* 5. In Figure #8, what is the item marked?   *Cavity*

# Chapter Ten

# INTERIOR FRAMING

The inspector should be familiar with the interior framing in a home. Even with solid masonry outer walls, the home has interior wood framing defining walls, ceilings, and floors.

**Guide Note**
*Pages 83 to 88 present information about the interior framing of a home. The inspection and reporting of interior walls, ceilings, and floors is covered in another of our guides — A Practial Guide to Inspecting Interiors, Insulation, Ventilation. Use these pages to learn about construction techniques.*

## Floors

Most homes have a **2-layer floor**, consisting of a subfloor and a finish floor above it. In balloon framing, the subfloor goes to the inner edge of the wall stud. In platform framing, the subfloor goes under the sole plate and to the exterior.

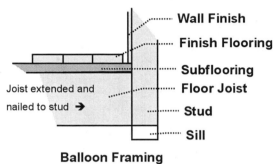

**Balloon Framing**

Joist extended and nailed to stud →
- Wall Finish
- Finish Flooring
- Subflooring
- Floor Joist
- Stud
- Sill

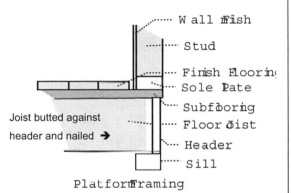

Platform Framing

Joist butted against header and nailed →
- Wall Finish
- Stud
- Finish Flooring
- Sole Pate
- Subflooring
- Floor Joist
- Header
- Sill

The subfloor today is normally **plywood, OSB, or particle board sheets 5/8" to 3/4" thick**, laid either perpendicular or diagonally to the joists. If the subfloor is laid on the perpendicular, the finish floor is then laid parallel to the joists. If the subfloor is on the diagonal, then the finish floor is laid either perpendicular or parallel to the joists.

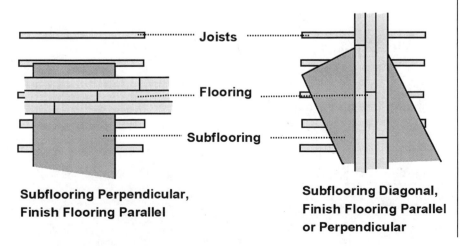

Joists
Flooring
Subflooring

**Subflooring Perpendicular, Finish Flooring Parallel**

**Subflooring Diagonal, Finish Flooring Parallel or Perpendicular**

The meeting edges of the plywood or particle board sheets are nailed to the joists. A structural adhesive may be applied to the joists to add strength to the connection.

Some subfloors are made with **tongue and groove planking**. The home inspector can determine whether this is true by examining the subfloor from the basement. In some cases, you may find that there is no subfloor under a hardwood finish floor. This is common in earlier buildings or in newer buildings where the builder just hasn't put a subfloor in under a hardwood floor.

The top layer of flooring may be one of the following:

- **Hardwood:** Tongue and groove strips from 1 3/4" to 2 1/4" wide, usually made of oak. But birch, beech, chestnut, maple, pecan, walnut, or hard pine can be used. Strip flooring is usually nailed. If no subfloor is put in, the strips should be at least 1 1/2" thick.

- **Softwood:** 1 x 4 tongue and groove strips, usually pine, although fir and cedar may be used.

- **Underlayment:** 1/4" plywood or particle board sheets are often used over the subfloor when the floor is expected to be finished with carpeting or resilient coverings such as linoleum or vinyl.

- **Ceramic tile, stone, marble, slate:** When these tiles are used, they are laid in a bed of mortar or adhesive on the subfloor or on a plywood underlayment. If no underlayment is used, the subflloor should be 3/4" thick to carry the weight of the tiles.

NOTE: Finish flooring may be applied to the concrete floor in slab on grade construction.

The problems we've already discussed relating to beams and joists (starting on page 54) can have a detrimental effect on the floor itself. Deterioration of girders and joists, cracking, sagging, twisting, rot, improper notches or cuts in these members, and loss of connection and/or support can all contribute to an abnormal condition in the floor. Here is an overview of signs the inspector may encounter in floors that indicate structural problems:

**Uneven floors:** A floor can have highs and lows in it. A **hollow** can be caused by the failure of a single joist. When a hollow is present in the floor along a partition, it may be

*Worksheet Answers (page 82)*

1. *A is Figure #3.*
   *B is Figure #2.*
   *C is Figure #4.*
   *D is Figure #8.*
   *E is Figure #5.*
   *F is Figure #1.*
   *G is Figure #6*
   *H is Figure #7.*
2. *Bracing (Girt)*
3. *Flashing*
4. *Ledger*
5. *Metal (brick) ties*
6. *Air space*

that the partition is built between the joists. When the hollow appears on either side of a doorway, it's an indication of poor support for the studs on either side of the opening. A **ridge** in an upstairs floor may be caused by a downstairs partition built parallel to a joist. There may be a **bulge** in the floor over a support column, indicating that the column is moving up or the house is moving down. Another cause of a bulge can be from an overloaded cantilevered joist, where the joist's interior end is being forced upward.

- **Unlevel floors:** Unlike uneven floors, an unlevel floor has a continuous slope in one direction. This can be caused by foundation settlement pulling the floor lower at the outer edges. The condition can also be caused by shrinkage of wood members, where interior walls will shrink more than the outer wood framing. In this case, floors are likely to tilt inward toward partitions.

- **Sagging floors:** This is where there is a low area in the middle of a room. This is largely due to overloading the floor without the proper supports being added to prevent the sag. Waterbeds, refrigerators, pianos, and other heavy objects can cause floor sag. More support is needed.

- **Deflecting floors:** These floors have upward and downward movement. **Bouncy floors** are usually due to weakness in the joists or a lack of proper bridging. **Soft** or **springy floors** can indicate a problem between the subfloor and joists — poor support of the subfloor by the joists because of poor nailing or loss of connection. Improper spanning are also causes of the above.

- **Noisy floors: Squeaks** in flooring are caused by a poor connection between the subfloor and the joists. Weight on the floor pushes the subfloor down to the joist, and the resulting squeak is caused by nails sliding in and out. The home inspector may notice **drumming** and **rattling sounds** from the floor. These sounds are associated with the joists, not the subfloor. Low frequency sounds can be caused by weak flexible floor joists. Higher frequencies are the result of stiffness in the joists.

---

**PROBLEM SIGNS**

- Uneven floors
- Unlevel floors
- Sagging floors
- Deflecting floors
- Noisy floors

---

*Guide Note*

*In slab on grade construction, the floor can indicate problems with the slab (see pages 70 and 71 of this guide).*

## A Word about Trusses

The home inspector may find floor trusses in a home. The floor truss is an engineered floor joist, usually made of 2 x 4's. The truss can span longer distances than the regular wood joist. The horizontal members of the truss are called **chords**; the inner members are called **webs**. Gusset plates, usually metal, act as connectors between the chords and webs. Vertical members appear in the truss at the ends, where the truss rests on the foundation wall, for example, and 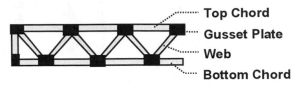 in interior spots such as above or below partitions for extra strength. Blocks may be supplied by the truss manufacturer to be inserted into the truss at particular points for additional support.

Some trusses are designed so that the top chord rests on the foundation wall. Others are designed for the bottom chord to rest on the wall. In these cases, installing the trusses wrong-side-up can seriously weaken their ability to carry the load for which they were intended. A tag should be present on the truss, indicating its *up* position.

The truss remains strong as long as it is properly installed, its connectors are securely attached to the members, no members of the truss are cut or removed, and the wood itself is not subjected to cracking or water and insect damage. Just as with wooden joists, problems with trusses can manifest themselves in the floor above. Floors can become uneven, sag, or deflect for the same reasons.

## Cantilevers

Joists or girders can be extended out to provide support for a deck or balcony, either inside or outside the house, without a support at the farthest end. This is called a **cantilever**.

A cantilevered structure can be springy if the joist or girder is not able to properly carry the load. This may be the case in older homes where no restrictions were placed on how much of the joist could extend. Today, it's recommended that only 1/6 of the length of the joist be unsupported.

With an **interior balcony**, as shown here, the downward load at the unsupported end of the joist is reflected by an upward load on the joist an equal distance from the support point. If the joist is overloaded

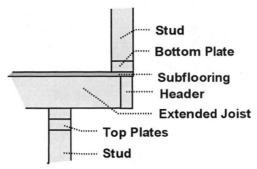

Stud

Bottom Plate

Subflooring

Header

Extended Joist

Top Plates

Stud

at its cantilevered end and is pushed downward, there can be a bulge in the floor at the other end of the joist. Or the joist can crush or crack where it is supported by the partition below. Sometimes, the uncantilevered portion of the room can be overloaded, causing the cantilevered area to rise.

With the cantilever that extends through the wall of the house to form an **outside deck or balcony**, these same problems can occur. When the cantilever is outside and exposed to the elements, the interface between the floor of the structure and the wall should be examined carefully for wood rot. This is an area that is particularly susceptible for leakage to occur.

## Other Framing Concerns

There are other structural framing concerns the home inspector needs to be aware of.

- **Load-bearing walls:** Some of the interior walls of a structure are load bearing while others are not. Ideally, a first-floor load-bearing wall should be constructed above the supporting beam in the basement and the second-floor load-bearing wall directly above so that the load from the roof is transferred vertically downward as shown here. But more often, the

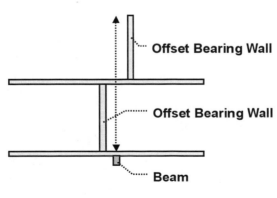

Offset Bearing Wall

Offset Bearing Wall

Beam

load-bearing walls are offset on either side of this vertical line. Generally a 3' offset is allowed if the load-bearing

*A cantilever is an extension of the floor structure which depends on the strength of the unsupported portion of the girder or joists to carry the load of the structure. A cantilever can be an interior balcony or an exterior balcony or deck.*

wall does not support a floor above it; otherwise only a 2' offset is allowable. An offset wall can cause the joists supporting it to be deflected with a resulting low spot in the floor at the wall and a hump over the beam or bearing wall below.

• **Double joists:** Joists that are subjected to concentrated loads may not be strong enough. Therefore, joists are doubled under non-load-bearing partitions to support this concentrated load. Joists should also be doubled where expected to support heavy live loads such as pianos, bookcases, and waterbeds.

NOTE: Load-bearing partitions should not rest on joists at all, but should be supported by beams.

• **Framing around stairs:** When an opening in the floor is needed, as around the opening for a stairway, joists are interrupted. These joists are secured to a header running perpendicular to the joists. The header carries the load from the joists over to the trimmer joists. When an opening is wide enough, the headers and trimmer joists are doubled for stability. Generally, if the opening is wider than 32", the trimmers are doubled. When the opening is 48" or more, the headers should also be doubled.

Headers and trimmers must be securely fastened to each other. Evidence of failing connections can be seen in the walls and ceiling around the stairway, where the members are pulling apart.

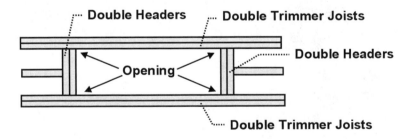

# Chapter Eleven

# INSPECTING ROOF STRUCTURE

This last chapter covers the inspection of the roof structure. The home inspector inspects the roof from the underside, using any information that might have been gathered from viewing the external roof. The inspection includes the following:

- Rafters, collar ties, and knee walls
- Trusses
- Roof sheathing
- Ceiling structure
- Attic leakage and condensation
- Insulation
- Ventilation

Most standards of practice for the inspection of the attic specifically state the following:

- Required to enter attic spaces except when access is obstructed, when entry could damage property, or when dangerous or adverse situations are suspected.
- Required to report the methods used to observe attics.

Just as with the crawl space, it's sometimes difficult to get into the attic to view the structural members. And sometimes it's not possible to walk through the attic and inspect the roof structure completely. So it's important for the home inspector to report the **methods used** to access the attic, letting customers know that there may be defects in the attic that the inspector was unable to find due to limited access. Sometimes, the attic may be finished and structural members may not be visible.

## Framing Members

From the attic, the home inspector will report on the condition of the roof structures, including rafters, collar ties, knee walls, and the ceiling joists.

**Rafters:** Rafters are structural members of the pitched roof, designed to support the roof sheathing and transmit the roof load to bearing walls and beams below. Rafters are made of wood 2 x 4's, 2 x 6's, or 2 x 8's, occurring every 16" to 24" along the length of the roof. In a flat roof,

*Pages 89 to 102 present the procedures on inspecting the roof structure. The inspection of the roof's coverings and other external aspects of the roof is covered in another of our guides — A Practical Guide to Inspecting Roofs. Insulation and ventilation are presented in greater detail in A Practical Guide to Inspecting Interiors, Insulation, Ventilation.*

*"A beginning home inspector
will walk on a ceiling only once
is his or her career. After
putting your foot through the
ceiling, you'll never try it again.
"Some attics have only the
ceiling joists — no actual floor
or walking surface. You
definitely don't want to step
between the joists to get a look
at the rafters. That's how you
break through the ceiling below.
Other attics may have a plank
walkway laid over the joists.
You'll want to be careful about
stepping on loose planks too.
"Be careful up there."*

*Roy Newcomer*

rafters are called roof
joists.

Rafters are placed on
edge, their narrow
dimension up for nailing.
The deeper dimension is
on the vertical to provide
strength. They are
connected to the outer
walls of the house in
various ways.

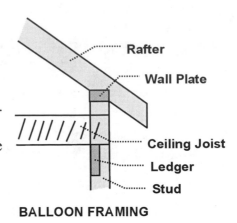

**BALLOON FRAMING**

In **balloon framing**, as shown above, rafters are notched
and connected to a wall plate that sits on top of the wall
studs. The rafters do not make contact with the ceiling
joists at all.

In **platform framing**, as shown below, rafters are attached
to the double top plate or a special rafter plate (illustration
on right). Where hurricanes are common, rafters are not
only nailed to the plate but securely attached with metal
ties. Rafters can be notched (left and right illustrations) or
end cut to sit flush on the plate (center). In roof framing,
there is a rafter for every ceiling joist. The rafter can be
nailed to the ceiling joist (left and center) or not (right),
depending on the design of the roof.

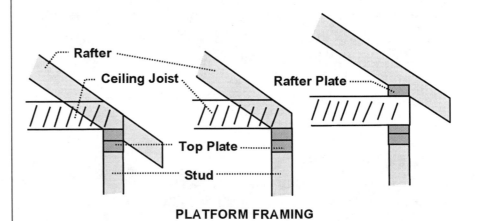

**PLATFORM FRAMING**

Rafters that are overloaded from several layers of roof
covering or ice and snow will **sag**. They can be supported
by collar ties or knee walls (see pages 92 and 93) which
absorb some of the roof load. Rafters can have a natural

curve or warp, known as a **crown**. During construction, carpenters will make sure that rafters with a curve will be installed with the crown facing upward. If rafters are installed with crowns facing both up and down, the result can be a **wavy** roof surface. The roof can also appear wavy if rafters are cut to unequal lengths. The home inspector can see sag and waviness from the exterior roof surface.

Rafters should be checked for **cracks**, warping, and sagging. All **rafter fastenings** should be secure and carefully examined — at ridge, plate, and joist connections and with supporting structures.

Rafters experience a phenomenon called **rafter spread**, where the roof load bearing on the rafters force them outward. This condition can be so bad that the soffit and the upper walls of the structure can be pushed outward along with the rafters. Secure connections to all framing members and collar ties and/or knee walls help prevent this problem.

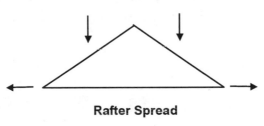

**Rafter Spread**

Rafters should be inspected carefully for **water** and **condensation damage** and should be probed where wood deterioration is suspected. Rafters should not be cut through.

**INSPECTING RAFTERS**

• Report any cracked or cut rafters.

• Watch for sagging, waviness, and rafter spread.

• Be sure rafters are securely fastened to all other framing members.

• Check for deterioration.

• When rafters support roof equipment, check for reinforcement.

**#47 Attic of an old brick house**

*Photo #47 shows the attic of an old brick house. You might be surprised to find a situation like this if you inspect old homes. These rafters are actually made of tree trunks, indicating just how very old the home is. This old roof has plank sheathing, which shows signs of deterioration. This condition was the result of leaking over several years.*

When equipment such as solar panels are mounted on the roof, the rafters below might have to be reinforced. This may be done with doubled or tripled rafters, a knee wall, or a vertical column. Where the supports rest on joists, the joists also should have the proper support.

- **The Ridge:** Rafters can be attached to each other where they meet at the peak of the roof or attached to a ridge board. In the conventional roof, the ridge is not a structural part of the framework. However, the ridge should not be twisted or cracked, and rafters should remain securely fastened to the ridge. If you'll remember, **Photo #10** presented a house that had a broken ridge board at the very center of the roof (although it's not easy to see in this photo). That house was parting in the middle and sinking on both ends. The point is, although the ridge is most likely not a *structural* member of the roof framework, its condition can tell you what's going on elsewhere.

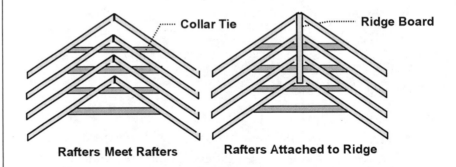

**Rafters Meet Rafters**   **Rafters Attached to Ridge**

- **Collar ties:** These are horizontal structural members of the roof framing system that prevent the rafters from sagging and from spreading. They are most often present in a steep sloped roof. They may form the framework for a finished ceiling in an attic room.

Collar ties are 2 x 4's, 2 x 6's or 1 x 8's and 1 x 10's attached securely to each pair of rafters about 1/3 to 1/2 down from the peak of the roof. The lower the ties, the more support they provide. There should be a collar tie for each pair of rafters, but you might see attics where there's a collar tie for every 2 or 3 pairs. Sometimes, if the roof load has increased since construction by adding several layers of roofing, one collar tie for every 2 or 3 rafter pairs is no longer enough, and collar ties should be added.

Collar ties are under **compression** from the weight of the roof pushing down on the rafters and under **tension** from the natural tendency of the rafters to spread outward. These forces can cause the collar ties to buckle, especially if the ties are 1 x 8's instead of 2 x 4's. Lateral bracing is usually used if collar ties are 8' long or longer.

When inspecting collar ties, the home inspector should look for buckling, deterioration of the wood, and for secure nailing to the rafters.

- **Knee walls:**  Knee walls are used with a low sloped roof to prevent the rafters from sagging.  They are small walls made of 2 x 4's that run from the attic floor to the rafters near their midpoint.  Knee walls may form the framework for side walls in a finished attic room. Knee walls must be properly secured to the rafters and ceiling joists to prevent movement.  Ceiling joists must be strong enough to support the wall.  Deflection of the joists beneath knee walls can be seen in damage to the ceiling finish in the rooms below.

In some cases, where the roof has a very long low slope, there may be both a knee wall at the lower ends of the rafters and collar ties on the higher ends.

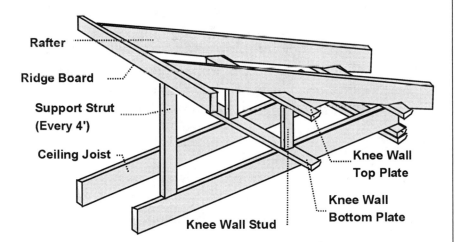

- **Ceiling joists:**  The joists in the attic support the attic floor and the ceiling of the rooms below.  In some cases, the home inspector may find 2 x 4 joists that were intended only to support the ceiling finish, not to provide an attic floor to be used as storage.  Joists must be securely fastened.  If overlapped or butted, they should be reinforced with a gusset plate to keep them from pulling apart at these junctions from the stress of rafter spread.

*Definitions*

*Rafters are the main structural members of the roof that support the roof covering and transmit roof loads to bearing walls or beams below.*

*The ridge is a horizontal framing board at the peak of the roof to which the rafters are attached.*

*Collar ties are horizontal structural members of the roof that connect opposite pairs of rafters together to prevent rafter sag and rafter spread.*

*Knee walls are supporting walls running from ceiling joists to rafters which prevent rafter sag.*

*Photo #48* shows an oddity that the home inspector may run into. This is the attic of a *prefabricated kit house* called a Lestrone house available in the 1940's. The exterior siding of the house is actually porcelain, and the roof is steel. When you see the attic of a Lestrone house, you may think that someone has jacked up failing rafters. However, this roof structure was designed and built to work with the jacks you see in the photo. The jacks run the entire length of the roof on both sides of the attic. Note the floor plates the jacks fit into.

#48 Attic of a prefrabricated kit house

### For Beginning Inspectors

*It's time to bother your friends again. This time ask to see the attic and take a look at rafter or truss construction. The more variations you see, the more you'll learn about roof structure and defects that may be present. It's probably important to point out at this time that you should refrain from scaring your friends about situations you're not sure about. During this learning process, it's best to let them know that you're not an expert yet. Of course, if you spot a situation that appears to be serious, you might suggest that someone else take a look at it.*

## Roof Trusses

A roof truss is an engineered geometric construction whose members perform the same function as rafters, collar ties, knee walls, and ceiling joists. Roof trusses used in residential homes are made up of 2 x 4's or 2 x 6's attached with wood or metal gusset plates nailed or glued in place. The bottom chord of the truss is used as the framework that supports the ceiling finish below. Trusses are normally placed every 24" the length of the roof.

Two commonly used trusses in residential construction are the **Fink truss** with the webbing and the **Howe truss** that incorporates vertical webs into its configuration.

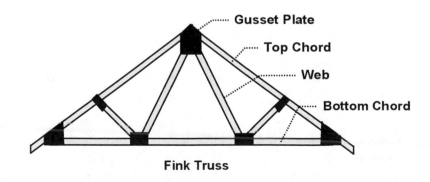

Gusset Plate
Top Chord
Web
Bottom Chord
**Fink Truss**

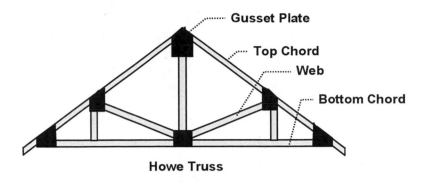

**Howe Truss**

There are many other truss designs with variations on the inner members, or webs, of the truss. Some are designed for heavy top chord loading or heavy bottom chord support. Some of the chords may be 2 x 5's or 2 x 6's. The **scissors truss**, as shown here, is used in vaulted or cathedral ceiling structures.

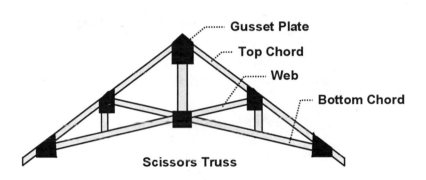

**Scissors Truss**

Trusses may require **permanent lateral bracing** to be added at the midpoint of some of the webs to keep the webs from buckling. Directions for installing the trusses may suggest that 1 x 4's be used for bracing (brightly colored tags are typically stapled to the webs where permanent lateral bracing is required). This partial view of a Howe truss shows where lateral bracing may be required.

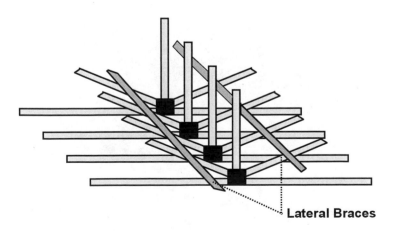

**Lateral Braces**

---

**INSPECTING TRUSSES**

- Watch for any cracked, cut, or missing truss members.

- Be sure bracing is present if it's required.

- Inspect gussets for corrosion or looseness.

- Check for deterioration.

- Be on the lookout for truss uplift.

- Look for bowing and warping.

When inspecting trusses, the home inspector must be sure to check a representative number of gusset plates to make sure that they are securely fastened. Look for loose nails, splitting of members at these connections, corrosion of the gusset plates, or broken plates. If the connections fail, the truss fails.

It is not always easy to get through an attic filled with truss webbing, but the home inspector should make every attempt to get down the length of the attic for a look. Watch for any water damage or deterioration of truss members. Be on the lookout for any cut or removed members of trusses. Trusses with missing members are not working trusses.

*Photo #49 shows a view looking up into the peak of a roof made up of **Fink trusses**. Here, the gusset plates are pulling away from the top chords and the webs. A closer look revealed that nails were pulling out . The plates needed to be refastened to prevent any movement of the trusses and further damage to the members.*

**#49 Fink trusses**

**#50 Howe trusses**

*Photo #50 is a view of an attic with **Howe trusses**. This attic had insulation blown in that covered many of the bottom chords of the trusses. If we'd just taken a look from the attic access area and not made our way into the attic, we would have been unable to read the red tag on the second web on the left in the photo. That tag said that lateral bracing was required along these diagonal webs. Of course, you can see that's it's not there at all.*

When the bottom chords of trusses are covered with insulation, as in Photo #50, a phenomenon called **truss uplift** can occur. What happens is that the bottom chords of the trusses deflect upwards as much as an inch or two during the winter weather. When the uplift is extreme, the truss can pick up the ceiling finish below, even separating the ceiling from the walls of the room. In some cases, the entire walls of the rooms below can be lifted, leaving a separation between the walls and the floor. If truss uplift is suspected, the home inspector should check the rooms below for these signs.

The experts believe that truss uplift is caused by temperature and humidity changes in the attic during the cold months. The bottom chord, buried in insulation and retaining more heat, reacts differently from the upper truss members. The bottom chord bows upward as a result. During the warmer months, the bottom chord moves downward again.

Truss uplift is not considered to be a structural problem, although it can cause cracks in the junction between the ceiling and wall below. One solution is to attach a molding in this room, where the molding is secured to the ceiling but not to the wall. Then when the ceiling moves up and down, the separation is not visible. Another solution is to disconnect the ceiling drywall from the truss and provide alternate ceiling support.

### Definitions

*A roof truss is an engineered, prefabricated geometric roof framing component. A chord is an outer member of the truss, either horizontal or diagonal. A web is one of the interior members of the truss, either vertical or diagonal. Gusset plates are metal or wood connectors that hold members of the truss together.*

*Truss uplift is a phenomenon where the bottom chord of a roof truss bows upward during the cold months and returns to its normal position during the warmer months.*

## Roof Sheathing

Roof sheathing, or roof decking, supports the roof covering and transmits its weight, as well as live weights such as snow and ice, to the rafters or trusses. It also serves to stabilize the roof structure by holding the rafters in position.

Roof sheathing before the 1960's was usually **planking**. Spaced planking is best for wood shingles and rigid roofing such as slate or tile, the spaces allowing for moisture exchange. Planking can also be butted or tongue and groove. There should be enough nails used to keep planks from warping and buckling. You'll be able to notice warped or buckled planking from the outside because of the visible waviness in the roof's surface.

**Plywood panels** for sheathing were introduced in the 1950's. **Waferboard, flakeboard panels, or OSB** have been around since the 1960's. Panels, generally 4' x 8' in size, are installed perpendicular to the rafters. The 4' edges should be nailed to the rafters, leaving 1/16" for swelling of the panel. The 8' edges are nailed to rafters and are secured to each other with metal H-clips between rafters. Plywood & OSB used today is usually 15/32" thick over rafters spaced at 24".

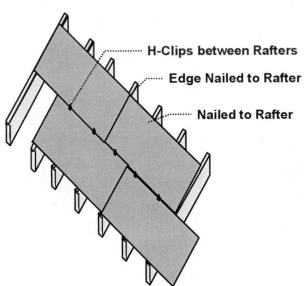

H-Clips between Rafters

Edge Nailed to Rafter

Nailed to Rafter

If the sheathing is too thin, it can sag under the roof load. Waviness in panel sheathing can be a sign that H-clips haven't been used (not required in all areas).

Panel sheathing should be carefully inspected for deterioration. Too much moisture and condensation in the attic can cause sheathing to delaminate. **Delamination** is a deterioration process in which the laminated, bonded layers in the sheathing will begin to separate. Most often, delaminated sheathing can be spotted by its color — it will turn darker or **black**, although this is not always the case.

*Definition*

*Delamination is a deterioration process during which the layers in the laminated plywood panel begin to separate.*

**#51 Delamination of the roof sheathing**

*Photo #51 shows the underside of a roof with serious **delamination of the roof sheathing**. The cause of this condition was the high moisture content in the attic.*

Delaminated plywood or OSB can no longer hold nails and can be dangerous to walk on.

## Water Penetration

The home inspector should examine the attic for any evidence of water leaking into the house. Examine the sheathing and rafters for any signs of deterioration from water penetrating through the roof surface. Inspect the chimney to see if there is any deterioration in the mortar and bricks from leaking flashings. Check carefully around other roof openings such as vents and piping to be sure they are water tight.

## Attic Moisture

Many homeowners misunderstand the principles behind attic insulation and ventilation and what effects they can have on the roof structure of a home. As a consequence of that misunderstanding, the home inspector may find insulation and ventilation in such a state that they actually promote attic moisture and resulting damage to the structural components.

When there is unlimited moisture in the attic air and no proper venting of the attic space, the wood members of the roof structure absorb this moisture and become saturated. Wood rot, mildew, and plywood delamination is the result. When moisture condenses inside the insulation, insulation rots and falls apart and wood members next to the insulation fall victim to rot and decay.

Here in an overview of the principles on insulation and ventilation:

- **Principle #1: The goal in the unfinished attic is to have the temperature the same as the outdoors.** Therefore, insulation should be laid on the attic floor, not between the rafters. When insulation is placed between the rafters, then the attic is part of the heated area of the house. Basically, the homeowner does not want warm air from the house to enter the attic space at all. Insulation laid between the ceiling joists acts as a heat retarder.

- **Principle #2: Warm air from the house brings moisture into the attic.** All attempts should be made to keep moisture and humidity from entering the attic space. An **air/vapor barrier** (a sheet of plastic such as polyethylene) should be laid beneath the floor insulation to enhance its ability to keep warm air from entering the attic space. If insulation has a **moisture barrier side**, it should be placed with this side down, toward the warm house below. Since warm air trapped in the insulation condenses as it cools, these barriers also protect the ceiling structure from the moisture that collects in the insulation.

  Another way moisture can enter the attic is if there are household fans that exhaust into the attic. The attic could be properly insulated and have an air/vapor barrier and then have plumbing vents and fan ducts pouring warm air into the attic. This would undo whatever good the insulation and barrier are doing. Plumbing vents and fan ducts *must* be vented directly to the outside.

- **Principle #3: The attic space must be vented to eliminate any moisture in the air.** It's not possible to completely prevent warm moist air from entering the attic from the house below. The cold side of the insulation needs to be ventilated to remove the moist air that does get through. The goal is to move this air to the outside.

  Homes are designed with various kinds of **attic venting systems** — gable vents, soffit vents, ridge vents, and/or vent fans in the roof surface or gables. Sometimes, homeowners will tape plastic over the vents in winter to "retain the heat." That's a serious mistake. Vents should be in operating condition at all times. Vent screens should not be blocked with debris or animal nests. And vents should

---

## ATTIC CONDITIONS

- The temperature in the unfinished attic should be the same as outdoors.

- Warm air should be prevented from entering the attic space from the house.

- Moist air in the attic should be vented to the outside.

- Plumbing vents and exhaust fans should vent to the outside, not into the attic.

---

not be covered with insulation. The purpose is to remove moist air from the attic, and this can't happen if the vents are obstructed.

The home inspector may find no signs of leaking in the roof and yet find rotting rafters, rusting nails and gusset plates on trusses, and delaminating plywood sheathing. This is the result of improper insulation and/or the lack of ventilation, which allow the attic air to become moisture laden. The home inspector finding these conditions should carefully investigate the cause of the problem.

**#52 Bathroom vent**

*Photo #52 shows a **bathroom vent** venting directly into the attic and pouring warm air against the peak of the roof. The sheathing in this attic is OSB, which are wood chips glued together. OSB doesn't "delaminate" as plywood sheathing does. However, when it becomes moisture laden, it begins to crumble. In this case, the OSB also started to turn black.*

It should be noted that the **finished attic** is a different story. When collar ties and knee walls define a heated attic room, the *attic space* becomes the area between the rafters and the finished room. There, the same principles apply. This space between the rafters and the room should be kept at the outside temperature. An air/vapor barrier should be laid up the outer side of the knee wall, and across the ceiling. Insulation should be installed along this same path with its moisture barrier side facing the room. A mistake often made is by installing insulation between the rafters in this type of attic space. But it should be on the ceiling and against the warm inner room. This type of attic space must also be ventilated to provide an escape for moist air collecting in the space.

## DON'T EVER MISS

- Cracked, bowing, or twisting rafters

- Rafter spread and sidewall separation

- Sags in the roof

- Delaminated plywood

- Loose truss fastenings, bowing trusses, cracked, cut or missing members.

- Deterioration of structural members

- Water penetration

## Reporting Your Findings

You'll probably have a page reserved for attic findings in your inspection report. That's where you should report on the roof's structure. Ideally, there should be sections on the page for reporting the following:

- **Access:** Don't forget reporting how the attic was observed, whether from the scuttle hole or from its interior. Note whether access was partial, complete, or not at all. Remind customers that you can't find defects in areas that you were not able to access. If you like, you can report where attic access is located — in the upper hallway, above the bedroom closet, and so on.

- **Rafters and trusses:** Identify the type of roof structure present. For rafters, note their size as 2 x 6, 2 x 8, or more. Then note their condition, not forgetting to note items from the Don't Ever Miss list at the right.

- **Sheathing:** Identify the type of sheathing present as plywood, solid planking, and so on. Be sure to record delamination in the sheathing, noting a ventilation condition as the cause if that's correct. Note other defects such as sagging and deterioration.

- **Ceiling structure (attic floor):** Note if trusses or wood joists are present. If the floor is finished, make a note to the effect that the underlying structure was not visible.

- **Water penetration:** Record any evidence of leaking into the attic and water stains on structural members or around the chimney and other places. Identify the location of such evidence. Don't ever miss reporting this!

- **Major repairs:** There can be roof structural problems serious enough to classify as major repairs in your inspection report. Certainly, **cracked rafters** and **delaminated plywood** fall into that category. If you've noted these problems on the attic page of your report, list them again as major repairs on a summary page at the back of your report.

# WORKSHEET

*Test yourself on the following questions.*
*Answers appear on page 104.*

1. How should plywood subflooring <u>not</u> be laid?
   A. Perpendicular to the floor joists
   B. Parallel to the joists
   C. Diagonally to the joists

2. What is the proper term to describe floors with a continuous slope in one direction?

   A. Deflecting
   B. Sagging
   C. Uneven
   D. Unlevel

3. Which condition will cause a hollow appearing in a floor?

   A. An overloaded cantilever joist
   B. A downstairs partition built parallel to the joists
   C. A partition on the same floor built between the joists
   D. Foundation settlement

4. Identify the parts of the truss that are lettered.

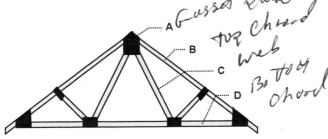
*(handwritten: A Gusset Plate, top Chord, web, Bottom Chord)*

5. Rafters may <u>not</u> be attached to:

   A. The top plate and the ceiling joists
   B. The ledger and wall studs
   C. A rafter plate only
   D. A wall plate only

6. H-clips are used to secure roof sheathing to:

   A. Roof sheathing
   B. Rafters

7. Identify the framing members of the roof structure shown here.

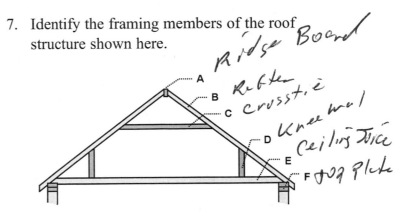
*(handwritten: Ridge Board, Rafter, crosstie, Knee wall, ceiling joice, top plate)*

8. What type of roof truss is presented in Photo #49?
   A. Fink truss
   B. Howe truss
   C. Scissors truss

9. What type of truss is presented in Photo #50?

   A. Fink truss
   B. Howe truss
   C. Scissors truss

10. Truss uplift is:

    A. A process where the layers of plywood begin to separate.
    B. A bowing <u>upward</u> of the <u>bottom</u> chord.
    C. Where the roof load pushes the rafters outward.
    D. Where bearing walls are offset from the supporting beam.

11. What is the function of collar ties?

    A. To support the attic floor and ceiling in the rooms below
    B. To support the roof sheathing
    C. To prevent the roof rafters from sagging and spreading
    D. To transmit the roof load to bearing walls
    E. To prevent ceiling joists from spreading

*Worksheet Answers* *(Page 103)*

1. B
2. D
3. C
4. A is a gusset plate.
   B is a top chord.
   C is a web.
   D is a bottom chord.
5. B
6. A
7. A is a ridge board.
   B is a rafter.
   C is a collar tie.
   D is a knee wall.
   E is a ceiling joist.
   F is a top plate.
8. A
9. B
10. B
11. C

# EXAM

A *Practical Guide to Inspecting Structure* has covered a lot of information. Now's the time to test yourself to see how well you've learned it. I included this exam in the guide so you'll have that chance, and I hope you'll try it.

*To receive Continuing Education Units:*
Complete the following exam by filling in the answer sheet found at the end of the exam. Return the answer sheet along with a $50.00 check or credit card information to:

American Home Inspectors Training Institute
N19 W24075 Riverwood Dr., Suite 200
Waukesha, WI 53188

*Please indicate on the answer sheet which organization you are seeking CEUs.*

It will be necessary to pass the exam with at least a 75% passing grade in order to receive CEUs.

*Roy Newcomer*

Name_____     Phone:_____

Address_____     e-mail:_____

_____     Credit Card #:_____

                                     Exp Date:_____

*Fill in the corresponding box on the answer sheet for each of the following questions.*

1. Which action is required by most standards of practice?

   A. Required to enter underfloor crawl spaces and attic spaces if the property has them.
   B. Required to probe structural components where deterioration is suspected.
   C. Required to report signs of repair work on the property.
   D. Required to report cracks in the foundation only if they are active cracks.

2. What is the overall purpose of the structural inspection?

   A. To report signs of water penetration
   B. To check for code compliance
   C. To recommend inspection by a structural engineer
   D. To identify major structural deficiencies

3. Which of the following is not part of the structural inspection?

   A. Foundation walls
   B. Roof framing members
   C. Interior finish materials
   D. Girders and floor joists

4. Vertical cracks in the foundation wall extending down to the footings can be caused by:

   A. Soil pressure against the wall
   B. Improper backfilling
   C. Settlement of the structure
   D. Overloading the lintel

5. For which type of foundation cracking should a structural engineer be recommended?

   A. Cracks of any width
   B. All active and inactive cracks
   C. Shrinkage cracks
   D. Horizontal cracks with movement

6. For which condition should the customer be advised to monitor the foundation wall?

   A. Cracks less than 1/4" wide showing little or no movement
   B. Step cracks indicating foundation settlement
   C. Excessive bowing of the wall with no cracks
   D. All active cracks

7. What sign indicates that a foundation crack is still active?

   A. Random cracking over the wall's surface
   B. Deteriorating mortar in the wall
   C. A pilaster constructed to support the wall
   D. Sharp edges to the crack

8. Water stains along the junction of the basement wall and floor most likely indicates:

   A. A leaking water heater
   B. A drain tile problem
   C. A high water table
   D. A faulty downspout or splash block

9. The source of most of the water leaking into basements can be attributed to:

   A. Surface water
   B. Ground water

10. Which of the following is a sign of water penetration through the foundation wall?

    A. A rotted girder or joist
    B. Efflorescence on the wall
    C. Water stains around the floor drain
    D. Delamination of the subfloor

11. What purpose does a Palmer valve serve?

    A. It allows the homeowner to turn on the sump pump manually.
    B. It provides a turnoff for an underground oil tank.
    C. It allows water to flow from the sump pump away from the house.
    D. It allows water to flow from the drain tile system into the sewer.

12. An end notch cut into a beam where it rests on the foundation may be up to 1/2 the depth of the beam.

    A. True
    B. False

13. Notches cut in the top of a joist may <u>not</u> be cut in the middle 1/3 length of the joist.

    A. True
    B. False

14. A member of a floor truss can be cut or removed only within 2' of the end of the truss.

    A. True
    B. False

15. When inspecting ledger strips in the basement ceiling structure, the home inspector should:

    A. Be sure gusset plates are in place and not rusting.
    B. Check their connection to support columns for crushing.
    C. Watch out for nails pulling out and for sagging and cracking.
    D. Look for cracked joists at the tenons.

16. Why should a dirt floor in a crawl space be covered?

   A. To prevent dust from entering the ducts and piping in the crawl space
   B. To prevent moisture from the soil being released into the crawl space
   C. To insulate the house above from cold released from the soil

17. Perimeter vents in a crawl space should be:

   A. Closed at all times.
   B. Open in winter and summer, closed in spring and fall.
   C. Open in spring and fall, closed in winter and summer.
   D. There is little agreement about when to keep vents open or closed.

18. Which of the conditions can be caused by poor ventilation in a crawl space?

   A. Deterioration of framing members
   B. Cracks in the foundation wall
   C. An uneven crawl space floor
   D. Water penetration in the foundation wall

19. In balloon framing construction, the wall studs rest directly on the:

   A. Foundation
   B. Sill
   C. Ledger

20. In platform framing construction the wall studs rest directly on the:

   A. Sill
   B. Header
   C. Sole plate

21. Which construction method is common in barns and churches?

   A. Balloon framing
   B. Platform framing
   C. Post and beam construction

22. Identify the components marked A,B,D,H, in this construction drawing below.

   A. Sill, stud, top plate, sole plate
   B. Header, bracing, top plate, sill
   C. Header, bracing, sole plate, sub floor

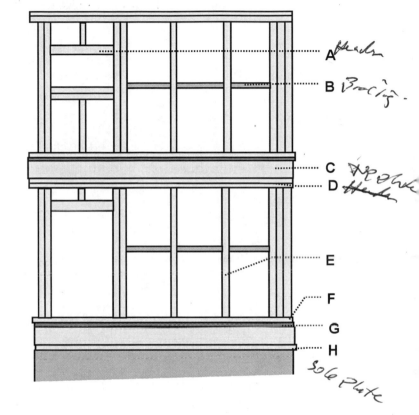

23. The wall framing method shown in the drawing above is an example of :

   A. Balloon framing
   B. Platform framing

24. In which type of framing does the subfloor go only to the inner edge of the wall stud?

   A. Balloon framing
   B. Platform framing

25. What would not be a cause of bulging or leaning in exterior masonry walls?

 A. Deterioration of the mortar
 B. Warping or expanding joists
 C. Rafter spread
 D. A rusting lintel

26. What should be suspected if a brick veneer wall has a bow to it?

 A. Loose brick ties
 B. Blocked weep holes
 C. Spalling
 D. Loose header rows

27. Step cracks above a window in a solid brick wall are most often caused by:

 A. A problem with the lintel
 B. Foundation settlement
 C. Loose brick ties
 D. Footing failure

28. Which can cause a floor to be bouncy?

 A. A downstairs partition built parallel to the joists
 B. A partition built between the joists
 C. Shrinkage of the outer wall
 D. Lack of proper bridging in the joists

29. Installing floor trusses wrong-side-up weakens their ability to carry their intended load.

 A. True
 B. False

30. What may happen to the floor as a result of a load-bearing wall being offset from the supporting beam below it?

 A. The joists supporting the wall may be deflected.
 B. A low spot may appear in the floor along the wall.
 C. A ridge may appear above the beam.
 D. All of the above.

31. Which statement is false?

 A. The home inspector is not required to enter the attic if access is obstructed.
 B. The home inspector is not required to enter the attic if entry will damage property.
 C. The home inspector is not required to enter the attic if dangerous conditions are suspected.
 D. The home inspector is not required to report the methods used to observe the attic.

32. What is the function of a knee wall in an attic?

 A. To support the roof sheathing
 B. To prevent the roof rafters from sagging
 C. To prevent the roof rafters from spreading
 D. To provide a surface for attaching insulation

33. Which of the following is generally not a structural part of the roof framework?

 A. Scissors truss
 B. Collar ties
 C. Ridge board
 D. Rafters

34. What condition should be reported as a major defect in the roof structure?

 A. Loose gusset plates in the trusses
 B. Truss uplift
 C. Cracked rafters and delaminated sheathing
 D. Waviness in the roof surface

35. What type of roof truss is used in vaulted or cathedral ceiling structures?

 A. Fink truss
 B. Howe truss
 C. Scissors truss

25.    C

26    A

27    A

28    D

29    A

30    D

31    D-

32    B

33    B

34    C

Arthr    35    C

36    Gusset Plate, Top Chord, Web.

37    Home Truss's   (B)

38   B   your truss —

39    D

40 -    A

41    C

42    B

43   (D)   A   Ridge Board, Rafter, Collar tie, Knee wall

44    A

~ Structure Test ~

1. A
2. D
3. C
4. C
5. D
6. D
7. D
8. B
9. A
10. B
11. D
12. B
13. A
14. B
15. B
16. B
?  17. C
18. A
19. B
20. C
21. C
22. A   Header
    B   Bracing
    D   Top Plate
    H   Sole Plate
23. B
24. A

36. Identify the members of the truss marked as A, B and C, in the drawing below.

   A. Web, top chord, gusset plate

   B.. Top chord, web, bottom chord

   C. Gusset plate, top chord, web

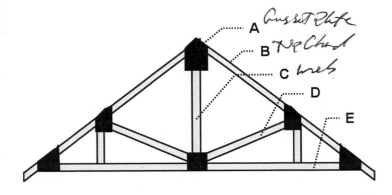

37. What type of roof truss is presented in the drawing above?

   A. Fink truss

   (B) Howe truss

   C. Scissors truss

38. What condition in the attic can cause plywood roof sheathing to delaminate?

   A. Too heavy live load on the roof

   (B.) Too much moisture in the attic

   C. No H-clips connecting the sheathing

   D. Warping and buckling

39. At what temperature should the unfinished attic be kept?

   A. Colder than outdoors

   B. Warmer than the house

   C. The same as the house

   (D.) The same as outdoors

40. When should attic vents be open?

   (A) At all times

   B. All seasons but winter

   C. Summer only

41. Where should insulation in an unfinished attic be laid?

   A. Between the rafters

   B. Behind the knee wall

   (C.) On the floor

   D. Over the collar ties

42. Insulation in the attic should be installed:

   A. With the moisture deterrent layer toward the attic space.

   (B) With the moisture deterrent layer toward the heated portion of the house.

   C. Either way.

43. Identify the framing members of the roof structure marked A, B, C and D, shown below.

   A. Ridge board, rafter, kneewall, collar tie

   B. Rafter, ridge board, collar tie, top plate

   C. Ceiling joist, ridge board, collar tie, kneewall

   D. Ridge board, rafter, collar tie, kneewall

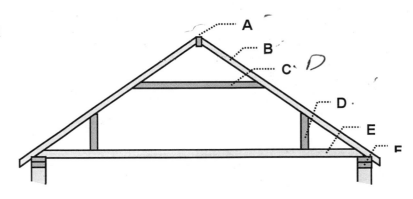

44. The home inspector should encourage the customer to:

   (A.) Follow along while the inspection is performed.

   B. Save their questions until after the inspection.

   C. Not purchase the house if a major defect is found.

45. At which communication step should the home inspector be particularly careful about not making any uneducated guesses?

A. Telling what is being inspected
B. Telling what is being looked for
C. Telling what is being done
D. Telling what is being found
E. Making suggestions about dealing with the findings

46. Why is it important to report storage areas in the basement?

A. To let the customer know the home inspector isn't snooping.
B. To let the customer know that storage areas are available.
C. To let the customer know that defects cannot be found in areas not visible.
D. To let the customer know the homeowners are hiding defects behind the storage area.

47. **Case Study:** You are inspecting the basement of a new home. Photo #20 shows the condition of the west foundation wall. A diagonal crack in the middle of the wall runs down to the footing and meets a vertical crack in the northwest corner. The crack is offset 3" and is pushing against the gas pipe at the top of the wall. The east wall is paneled. The south wall has random surface cracking. When you try the sump pump, you find it won't operate even when you pull up the lever.

How do you identify the foundation in the inspection report?

A. Poured concrete
B. Block
C. Stone
D. Wood
E. Brick

48. For the case study, which wall if any would you report as not visible?

A. The north wall
B. The south wall
C. The east wall
D. The west wall

49. For the case study, how would you report basement moisture (refer to Photo #20)?

A. Water stains are normal
B. Water stains at wall and floor need to be further evaluated to determine the cause
C. Do not report them
D. Have concrete floor removed to determine the cause

50. For the case study, what would you report as a major repair in the inspection report?

A. Offset diagonal crack on west wall. Foundation wall pushing on gas pipe. Sump pump not operating.
B. Offset diagonal crack on west wall. Foundation wall pushing on gas pipe. Random cracking on south wall
C. Random cracking on south wall. Foundation wall pushing on gas pipe. Sump pump not operating.
D. Offset diagonal crack on west wall. Random cracking on south wall. Sump pump not operating.

# GLOSSARY

**Angled crack** A diagonal foundation crack caused when the upload and download on the foundation offset each other.

**Balloon framing** A construction method where long vertical studs and corner posts run from the foundation to the roof and the floors are hung on the wall frame.

**Beams** See *Girders*.

**Below grade** Below the surface of the ground.

**Bending** In terms of structure, the movement of a structural member out of its original position without shearing as a result of forces applied to the member.

**Bleeder** Piping that runs from exterior drain tile under the footings to interior tile and to the sump pump.

**Blocking** A bracing method using a brace of wood between joists that is the same depth as the joists giving stiffness to the joists.

**Bow** A vertical curve in a wall, where the wall has an outward curve from top to bottom.

**Brick ties** Accordion-style metal fasteners used to attached a brick veneer to the wood framework of a home.

**Brick veneer** A wall construction method in which an outer layer of bricks is attached to the wood framework of the house using brick ties.

**Bridging** A bracing method using a 2 x 4 brace of wood between joists giving stiffness to the joists. Cross bridging uses diagonal 1 x 3's between joists.

**Bulge** A combination of both bow and sweep in a wall.

**Cantilever** An extension of the floor structure which depends on the strength of the unsupported portion of the girder or joists to carry the load of the structure. Can be an interior balcony or an exterior balcony or deck.

**Cavity wall** A masonry wall with a dead air space left between inner and outer layers of masonry.

**Chinking** A mortar made of clay, sand, and other binders such as animal hair used to fill the gaps between logs in a log home.

**Chord** An outside member of a truss, horizontal or on an angle.

**Cinder block** A block made of slag from steel making or railroad cinders that were used in home construction.

**Collar ties** Horizontal structural members of the roof that connect opposite pairs of rafters together to prevent rafter sag and rafter spread.

**Columns** Vertical supports that carry the weight of the structure from the girders to the ground, transmitting weight to the footings.

**Compound wall** A solid masonry wall built of two different materials.

**Compression** A stress that pushes on a structural member tending to make that member reduce its size.

**Concrete block** A block made of crushed stone and builders sand used in construction, especially foundation walls.

**Concrete masonry unit** Various kinds of hollow-core blocks used in foundation construction. Abbreviation *CMU*.

**Crawl space** The unfinished area between the house floor and the ground.

**Cross bridging** See *Bridging*.

**Dead load** The weight of a structure itself, its sheathing and wall coverings, and other integral components.

**Delamination** A deterioration process during which the layers in the laminated plywood panel begin to separate.

**Drain tile** Flexible, perforated plastic piping used in footing drains.

**Dry laid** A stone foundation laid stone upon stone without mortar or footings.

**Efflorescence** The white mineral deposit left after water passes through the foundation wall bringing dissolved salts from the wall material. It appears on the interior wall after the water has evaporated.

**Finish flooring** The flooring applied over the subfloor that provides the finish floor for the home.

**Fink truss** A triangular roof truss with w-shaped interior webbing.

**Flashing** For a foundation, a sheet metal placed over the foundation to prevent water leaking into the foundation.

**Floating slab** Where an independent slab does not rest on the foundation wall.

**Floor truss** An engineered, prefabricated rectangular floor framing component.

**Footing drain** A drainage system laid around the perimeter of a foundation below the level of the slab which drains water from the soil to another location.

**Footings** The bases on which the foundation rests which support and distribute the weight of the structure to the soil.

**Foundation** The part of the structure that supports it, transmits its weight from above-grade walls to the footings, and protects it from effects of soil pressure.

**Frost line** The depth of penetration of frost into the ground, normally about 4' below grade in cold climates.

**Girders** Horizontal load-bearing members of a floor system that carry the weight of the floor and wall loads to the foundation and columns. Also called beams.

**Grade** The slope of the land at the building site.

**Grade beams** Poured reinforced concrete beams that rest on grade, just in the ground, or on piers.

**Gusset plate** A piece of wood placed over partial bearing joists and nailed into the joists to hold them in place. In a truss, metal or wood connectors that hold members of the truss together.

**Header joist** The perimeter joist nailed to the sill. Also called the rim joist.

**Header rows** Rows of brick turned small end out as ties to hold a brick wall together.

**Horizontal crack** A horizontal foundation crack often caused by pressure being applied by soil outside the foundation.

**Howe truss** A triangular roof truss that incorporates vertical webbing into its configuration.

**Joists** Horizontal members of a floor system that carry the weight of the floor to the foundation, girders, or load-bearing walls.

**Knee walls** Supporting walls running from the ceiling joists to rafters which prevent rafter sag.

**Ledger** In balloon framing, a horizontal framing member attached to the wall studs providing supports for 2nd-floor joists.

**Ledger strips** Lengths of wood nailed along the bottom edge of a girder to provide support for joists.

**Lintel** A horizontal framing member that carries the load above a window or door opening.

**Live load** The weight of the home's occupants, the furnishings of the home, and other weight the structure must support.

**Monolithic slab-on-grade** Construction where the slab and the foundation are poured as one piece.

**Mortar** A mixture of a binder (lime, masonry cement, portland cement), an aggregate (sand), and water. Used to bond masonry units such as concrete blocks, stones, and bricks together.

**Mortise and tenon connections** Joining joists to girders where a mortise (hole) in the girder accepts the tenon (a projecting end) of the joist.

**Palmer valve** A hinged valve in the floor drain that allows water from the drain tile system to flow through the floor drain into the sewer.

**Pedestal-style sump pump** A pump which has the motor mounted on a shaft that sits above the water level.

**Piers** Columns supporting a structure that are built on footings in a hole below the frost line.

**Pilaster** A masonry column built against a wall to help absorb the horizontal load and stiffen the wall.

**Piles** Columns supporting a structure that are driven into the ground to reach soil of bearing strength.

**Platform framing** A construction method where the stories of the house are constructed one on top of each other.

**Post and beam** An old construction method in which a small number of posts and beams carry the weight of the structure to a plank flooring.

**Poured concrete** In a foundation wall, concrete poured into forms that are removed after the concrete has set, thereby forming a wall.

**Pressure-treated wood** Wood impregnated with chemical preservatives that protect the wood from termites and fungi that cause rot.

**Racking** A condition where a structure leans in such a way that the angles it forms with the foundation are no longer 90°.

**Rafters** Structural members of a pitched roof that support the roof covering and transmit roof loads to bearing walls and beams below.

**Rafter spread** A phenomenon where the roof load bearing on the rafters forces them outwards.

**Resting slab** Where a slab is laid to rest on top of a conventional foundation.

**Ridge board** A horizontal framing board at the peak of the roof to which rafters are attached.

**Roof truss** An engineered, prefabricated geometric roof framing component.

**Scissors truss**  A roof truss used in vaulted or cathedral ceilings.

**Shear**  A stress resulting from forces being applied upon a structural member from opposite directions.  Can cause the member to crack, split, or completely separate.

**Sheathing**  Sheets of plywood or wood planking used to cover a roof, wall, or floor frame.

**Sill**  The 2 x 4 or 2 x 6 laid flat on and anchored to the foundation, providing a pad for the framing system.  Also called a sill plate.

**Slab-on-grade**  A poured concrete slab that rests directly on the ground.

**Sole plate**  In platform framing, the horizontal framing member nailed to the header and floor joists at the outer edge of the structure.

**Spalling**  The crumbling and falling away of the surface of bricks, blocks, or concrete.

**Span**  The distance center to center between two like framing members such as floor joists.

**Step crack**  An angled crack appearing along the block joints in a concrete block foundation.

**Structure**  A home's skeleton, including its foundation and footings, roof, and framework.

**Subflooring**  Horizontal floor members that transfer the load of the home's furnishings and people to the floor joists.

**Submersible sump pump**  A pump that sits below the water level in the sump.

**Sump pump**  A pit located in the basement floor containing an electric pump that pumps water from the perimeter drain system away from the house.

**Supported slab**  Where the edges of a slab rest on a ledge at the top of the foundation wall.

**Sweep**  A horizontal curve in a wall, where the wall has an outward curve from side to side.

**Tension**  A stress that pulls at a structural member tending to make that member increase in size.

**Top plate**  In platform framing, the horizontal framing member nailed to the wall studs at the top of each story.

**Truss**  An engineered, prefabricated framing member.

**Truss uplift**  A phenomenon where the bottom chord of a roof truss bows upward during the cold months and returns to normal position during the warmer months.

**V-crack**  A crack that increases in width along its path.

**Vertical crack**  A vertical foundation crack often caused by settlement of the structure.

**Wall studs**  Vertical wall framing members. Connected in platform framing to the sole plate and the top to a top plate.  In balloon framing, connected to the sill.

**Web**  An interior member of a truss.

**Weep holes**  Openings in the bottom row of brick in a veneer wall providing an exit for water accumulating behind the veneer.

# INDEX

# A Practical Guide to Inspecting Program
### Study Unit One, Inspecting Structure

Student Name: _____ Date: _____

Address: _____

Phone: _____ Email: _____

Organization obtaining CEUs for: _____ Credit Card Info: _____

After you have completed the exam, mail *this exam answer page* to American Home Inspectors Training Institute. You may also fax in your answer sheet. You will be notified of your exam results.

Fill in the box(s) for the correct answer for each of the following questions:

| | | |
|---|---|---|
| 1. A☐ B☐ C☐ D☐ | 24. A☐ B☐ | 47. A☐ B☐ C☐ D☐ E☐ |
| 2. A☐ B☐ C☐ D☐ | 25. A☐ B☐ C☐ D☐ | |
| 3. A☐ B☐ C☐ D☐ | 26. A☐ B☐ C☐ D☐ | 48. A☐ B☐ C☐ D☐ |
| 4. A☐ B☐ C☐ D☐ | 27. A☐ B☐ C☐ D☐ | 49. A☐ B☐ C☐ D☐ E☐ |
| 5. A☐ B☐ C☐ D☐ | 28. A☐ B☐ C☐ D☐ | 50. A☐ B☐ C☐ D☐ |
| 6. A☐ B☐ C☐ D☐ | 29. A☐ B☐ | |
| 7. A☐ B☐ C☐ D☐ | 30. A☐ B☐ C☐ D☐ | |
| 8. A☐ B☐ C☐ D☐ | 31. A☐ B☐ C☐ D☐ | |
| 9. A☐ B☐ | 32. A☐ B☐ C☐ D☐ | |
| 10. A☐ B☐ C☐ D☐ | 33. A☐ B☐ C☐ D☐ | |
| 11. A☐ B☐ C☐ D☐ | 34. A☐ B☐ C☐ D☐ | |
| 12. A☐ B☐ | 35. A☐ B☐ C☐ | |
| 13. A☐ B☐ | 36. A☐ B☐ C☐ | |
| 14. A☐ B☐ | 37. A☐ B☐ C☐ | |
| 15. A☐ B☐ C☐ D☐ | 38. A☐ B☐ C☐ D☐ | |
| 16. A☐ B☐ C☐ | 39. A☐ B☐ C☐ D☐ | |
| 17. A☐ B☐ C☐ D☐ | 40. A☐ B☐ C☐ | |
| 18. A☐ B☐ C☐ D☐ | 41. A☐ B☐ C☐ D☐ | |
| 19. A☐ B☐ C☐ | 42. A☐ B☐ C☐ | |
| 20. A☐ B☐ C☐ | 43. A☐ B☐ C☐ D☐ | |
| 21. A☐ B☐ C☐ | 44. A☐ B☐ C☐ | |
| 22. A☐ B☐ C☐ | 45. A☐ B☐ C☐ D☐ E☐ | |
| 23. A☐ B☐ | 46. A☐ B☐ C☐ D☐ | |